T0136428

About Island Press

Since 1984, the nonprofit organization Island Press has been stimulating, shaping, and communicating ideas that are essential for solving environmental problems worldwide. With more than 1,000 titles in print and some 30 new releases each year, we are the nation's leading publisher on environmental issues. We identify innovative thinkers and emerging trends in the environmental field. We work with world-renowned experts and authors to develop cross-disciplinary solutions to environmental challenges.

Island Press designs and executes educational campaigns, in conjunction with our authors, to communicate their critical messages in print, in person, and online using the latest technologies, innovative programs, and the media. Our goal is to reach targeted audiences—scientists, policy makers, environmental advocates, urban planners, the media, and concerned citizens—with information that can be used to create the framework for long-term ecological health and human well-being.

Island Press gratefully acknowledges major support from The Bobolink Foundation, Caldera Foundation, The Curtis and Edith Munson Foundation, The Forrest C. and Frances H. Lattner Foundation, The JPB Foundation, The Kresge Foundation, The Summit Charitable Foundation, Inc., and many other generous organizations and individuals.

The opinions expressed in this book are those of the author(s) and do not necessarily reflect the views of our supporters.

Rural Renaissance

Rural Renaissance

REVITALIZING AMERICA'S HOMETOWNS
THROUGH CLEAN POWER

L. Michelle Moore

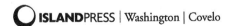 ISLANDPRESS | Washington | Covelo

Library of Congress Control Number: 2021950888

All Island Press books are printed on environmentally responsible materials.

Manufactured in the United States of America
10 9 8 7 6 5 4 3 2 1

Keywords: agrivoltaics, broadband, clean power, cooperatives, Department of Energy,
electric grid, electric vehicles (EVs), energy democracy, energy efficiency, energy
financing, energy justice, energy policy, energy storage, microgrid, the New Deal,
public power utilities, resilience, rural America, rural electric cooperatives, Rural
Electrification Administration (REA), smart grid, solar panels, solar power, US
Department of Agriculture (USDA), utilities, wind power

To Curtis Wynn, my dear friend, respected mentor,
and visionary rural cooperative leader.

As CEO of Roanoke Electric Cooperative, Curtis built the utility of the future in Eastern North Carolina. The rural community he served has energy efficiency to save on the bills, community solar, energy storage, a restorative program that sites solar on Black-family-owned farms, electric vehicles, vehicle-to-grid charging, and, increasingly, affordable broadband.

Curtis's influence extends beyond North Carolina. He is a national leader with four decades in the rural cooperative community, which includes serving as the National Rural Electric Cooperative Association's elected board president. His ability to pair gentle encouragement with a bold challenge led the NRECA to take the biggest step forward in governance that it has taken since its founding—adopting a national diversity, equity, and inclusion policy resolution in March 2021. Curtis now serves as CEO of SECO Energy, a two-hundred-thousand-member electric cooperative utility in Central Florida that is among the ten largest cooperative utilities in the country.

With gratitude, *Rural Renaissance* is dedicated to Curtis.

Contents

Prologue

I grew up in a small town in rural Georgia that I could not wait to escape. My worst nightmare was to end up back there, working for a textile mill—but that's exactly what happened, putting me on a path that shaped the rest of my life in ways I could never have imagined. From LaGrange, Georgia, to the White House and back again, and around the world in between, going home and falling back in love with the little town where I grew up has shown me how we can make every hometown in America a place where people can thrive. The urbanization of America (and the globalization of the world) doesn't represent the one true path to economic prosperity and a sustainable way of life. We can live small and live well too. And we can do it together.

Rural Renaissance is a story about power and empowerment. There's a moral to the story, and there are heroes in it, too, including an unlikely few who never envisioned the far-reaching good they would do. Most important, there's a happy ending, but only if we choose to pursue it.

It's also a story about energy—the lifeblood of opportunity and economic development and a pillar of human civilization in this twenty-first-century world. Turning on the lights in 1930s rural America welcomed even the farthest-flung communities into the industrial age and created an economic engine that attracted jobs and supported a quality of life previously unimaginable. In the age of information and the Anthropocene period, the challenges are different. Our old energy choices are destroying the places we love, and the places many of us come from are fading away. But we can be better.

Let me share a different vision, one that's possible thanks to some prescient decisions from our past: a future in which rural communities are flourishing. In this future, energy—clean, resilient, and local—and the infrastructure investment it takes to build it are sustaining good jobs in small towns, using the land rural America has in abundance to feed big-city energy appetites. Farming families are using unplanted fields for the "last crop," earning a living from the power that's freely and abundantly available in the form of solar energy, thereby preserving generational wealth. And communities are becoming more resilient, connecting local renewables with energy storage to help improve energy futures for everyone.

Built and financed alongside the modern and secure power grid that needs it to function, broadband runs to every home, and the network is owned and governed by the communities it serves through rural and small-town public power utilities. These new high-speed networks also abound with educational options for kids who can barely get a dial-up connection and a dog-eared textbook today. Everyone has remote access to doctors, and no one has to drive hours for basic health care. Local businesses and start-ups can succeed because they're finally connected to a global marketplace of people and companies that want to buy what they have to sell.

Rural and small-town utilities, the economic hearts of their hometowns, are flourishing because clean energy and the electrification of the transportation sector are enabling them to grow again. More revenue means more funding to support community development and pay for essential services, so residents are thriving, too, because their utilities represent their needs and are governed in alignment with their democratic roots.

As a result, and perhaps most important, the 48 percent of Americans whose hearts are in the country and who want to live in a small town can have a good life there with the dignity of good jobs, affordable homes, and the hope of an even better life for their kids.[1]

How can we make this vision a reality? America's nearly three thousand rural electric cooperatives and public power utilities were designed to be energy democracies. Founded to deliver power to the people who'd been left behind by the industrialization of American cities, these public purpose utilities were created as locally owned and governed institutions to advance economic opportunity and a better quality of life. Today, they serve more than 90 percent of the persistent poverty counties in the country and more than half of America's landmass—and they hold the legal authority to deliver clean energy and, with it, broadband that would reconnect rural America to the global economy in a way that enriches the places they serve rather than impoverishing them. As long as they are democratically governed in accordance with their mission, these local utilities can help build clean energy futures in which we can all thrive.

Seems a bit far-fetched? It's not—and in fact these opportunities have already begun to take root in places like Ahoskie, North Carolina; Ouachita, Arkansas; and my hometown of LaGrange, Georgia. Building a better future for rural America starts with celebrating and scaling what we're already doing right.

This story is my story too.

Mammaw Knopp's house stands at the end of Cleveland Drive on the edge of the Dunson Mill Village in LaGrange, Georgia. I lived there with her, Pappaw, and my aunt Ida much of my first-grade year. There was no air-conditioning in the summertime and no heat in the bedrooms in winter. We kept cool with a window fan and warm under three generations of hand-pieced quilts and hand-me-down blankets. And it was heaven.

That was in 1977, when the textile mills where my grandparents had labored their whole lives were beginning to fail. A fifty-five-year-old child of a sharecropper couldn't compete with cheap labor overseas, and the mills were winding down. Just ten years later, they'd be hollowed out, their beautiful red brick and heart pine bones waiting quietly for fresher eyes to find the beauty in them and bring them back to life.

If the textile mills were the job-generating brawn of my hometown, the utility was its brains. Established in the 1920s by state law and local philanthropy, the LaGrange municipal utility was created to bring power, light, and the prosperity that came with them to every local home at a time when the state's big private power company couldn't be bothered to serve the community. Over time, like many other small utilities across

Mammaw and Pappaw Knopp's pay stubs from the cotton mills, 1966. (Photo by author.)

the country, it became a central feature of life in LaGrange. Managing electricity, water, gas, and trash, it met the basic needs of the people and businesses it served. More than that, profits from the utility helped pay for the fundamental public service functions of a local government, funded by its central enterprise. LaGrange had no property taxes, but you'd better pay your light bill on time.

That same utility would send my mammaw and pappaw a $300 bill when winter temperatures got cold enough to freeze the pipes and they'd have to turn on their ancient furnace. The thermostat was set to barely 50 degrees, but we might as well have been burning money to keep the pipes warm. A bill like that could break the budget, but it wasn't because the energy itself was expensive. My mammaw's house was the only place I ever thought of as home, but there was not an inch of insulation in it, and the major appliances were all as old as I was. It was a loving home, but not energy efficient, which is why we got those backbreaking power bills.

Mammaw and Pappaw got by, but there were many people in our community who suffered. My bus route to what was then Troup Junior High School took us down Hammett Line Road just before we turned onto the school's driveway. What I can only describe as shacks lined either side of the route. There were outhouses in the back, chimneys for heat, and I always wondered if those homes had light or running water. The families who lived there mattered just as much as any other family in our town, but their economic poverty was aggravated by an energy poverty that made their lives even harder.

That's how it was and still is for many, but it doesn't have to be that way. Instead, we can be inspired by the thousand-acre Butler Solar Farm in Taylor County: Route 96 from Columbus to Macon runs straight through it, with solar on the right side and farming on the left. We can replicate the pioneering energy equity programs that have municipal utilities investing in energy and water efficiency to cut bills and improve housing—prioritizing the people who most need savings and directing reparative investments into communities. Our future can be focused on investing in people so that the jobs, businesses, and investors that are delivering our energy are homegrown and look like America. And it can be sustained by local energy economies that would have made sure our neighbors on Hammett Line Road had power and a place to live that respected their dignity too.

The only thing we need to do to build better futures is to make the choice. We don't need permission. We don't need anyone else to do it for us. There's plenty of money to do it, and it pays for itself in time. Clean energy provides the technology, its economics work, and it specifically does not take an act of Congress. (They did that back in 1936.)

Why now? Or perhaps more to the point, why not yet?

The climate and clean energy movement has been obsessed with "the City." It made mathematical sense. The urban core of America is home to more than 30 percent of the population, and urban activities generate approximately 75 percent of America's climate pollution.[2] If rural America was considered at all, it was as a big outdoor museum that happened to grow farm-to-table food. Because the climate and clean energy movement focused on policy and pollution reduction as key performance indicators, it was easy for rural America to get overlooked. As a result, big philanthropy and government poured money into cities for "green" buildings, clean transit, electric bus fleets, and other programs. While the quality of life in cities got a little better for some, clean energy and climate action became the markers of an affluent urban lifestyle that left everyone else out. Urban climate policy reduced pollution, but it didn't result in better lives for enough people.

In fact, at the same time, the quality of life in rural communities was and has remained in decline. In rural America, mortality rates are up and people are dying more often from preventable disease than they do in cities.[3] Deaths from opioid overdoses, alcoholism, violence, and suicide have also increased.[4] At the same time, rural hospitals are closing—136 since 2010, and 20 in 2020 alone.[5] Where rural hospitals remain, on average it takes twice as long to get to them as in urban communities.[6]

Education has suffered along with health. While rural students score better and graduate high school at higher rates that their urban counterparts, they attend and finish college at much lower rates.[7] Telemedicine and remote learning might improve outcomes, but nearly one-fifth of rural residents cannot access broadband.[8] As I have observed in my own community of friends and family, many rural residents are consumed with mistrust, rage, and grief.

This stark divide between urban and rural communities is in sharp contrast to their growing interdependence, as Daniel T. Lichter and James P. Ziliak observed in "The Rural-Urban Interface: New Patterns of Spatial Interdependence and Inequality in America." This breach extends from agriculture and technology to culture and design. To repair it, we've got to appreciate how urban and rural communities need each other and act on the fact that our futures and respective well-being are fundamentally intertwined.[9]

Energy systems are a critically important part of the picture, and they are part of the solution too. Not only do urban centers need the energy that rural communities can produce, but the economic, environmental, and human impacts that our energy systems generate also shape rural wealth, health, affordability, and quality of life. Shifting away from industrial energy systems that take resources from rural communities, burn them, and export the energy (leaving behind the pollution) and toward localized energy systems that are clean and abundant can be a powerful approach to renewing rural quality of life.

Whether we're talking about energy, water, or food—or just a place to get away and enjoy the restorative qualities of the natural world—big cities and small towns are bound together in a "network of mutuality" that is as beautiful as it is inescapable. I, for one, would never want to leave.

Though, once upon a time and for about twenty years, I tried. It was only while working for President Obama that I realized my entire career up to that point had saved a whole lot of big companies a whole lot of energy and had helped a few wealthy homeowners live a little greener—but hadn't done anything that would have directly helped my grandparents, though they always told me they were proud.

In 2015, I went home again to put what I knew to work for the people and the places I held most dear. This is what I learned.

Introduction

This book is about revitalizing our hometowns with clean power to create what I envision as a "rural renaissance." It is intended as a resource for people who want to help lead their communities toward a thriving clean energy future. The chapters that follow contain history, information, road maps, and examples relevant to that goal, with some real-world inspiration along the way.

Clean power makes this rural renaissance possible because clean energy technology can be localized. Instead of big power plants and long transmission lines that extract wealth from entire regions, we can have local solar projects that invest in neighborhoods. Add in energy storage and microgrids for resilience, electric vehicles to expand the opportunities, and energy efficiency for affordability and comfort, and we've got a great set of tools to help the places we live to thrive.

Importantly, localized clean energy systems are also aligned with the federalist nature of how we make decisions about energy and the local diversity of our ecoregions. This integration of natural resources, governance, and technology empowers us to create *local* energy futures based on what our communities need and what we have to share.

It was different in the past. In the nineteenth century, when our current energy system was built, we depended on industrial-sized power plants that burned fossil fuels dug out of the ground to make electricity, which had to be transported over hundreds of miles of overhead transmission lines to keep the lights on. Even if you could get your power locally, mines, fracking operations, oil refineries, and coal-fired power plants didn't make good neighbors. The amount of capital required to extract, generate, and transmit electricity also meant that only a few big companies and wealthy investors were able to own the power. No one else could afford the necessary investments, though everyone got stuck with a utility bill and the public cost of pollution. Clean power changes the equation by enabling our energy systems to move from an industrial scale to a local, and much more human, one.

Now is the time to make this future happen. Like much of the world, America is modernizing and rebuilding its energy infrastructure at a rate of investment unseen since the New Deal. Leveraging that momentum, and that money, to revitalize our hometowns with clean energy is an opportunity that's too good—and too rare—to miss.

It won't happen without you.

The times are ripe with transformational potential, but only if we examine not just how we got here, but why. Systems produce outcomes according to the values upon which they are founded. We can tinker at the edges for a lifetime, trying to change the mechanics of a system, but focusing on how a system works won't, and truly can't, produce lasting change. For transformation, we've got to go deeper, to the values level, and ask why. If we fail to question and transform the values at the center of our energy systems, we risk creating a clean energy dystopia that's powered by renewable resources but otherwise replicates the health and wealth disparities of the past.

As we think about how to undertake this work, here are seven principles that form the foundation of a rural renaissance and recur throughout the book:

Align value with values. Or as I learned in Sunday school, "Where your treasure is, there your heart will be also."[1] Currently, the solar tax credit (or ITC), incentivizes new solar energy systems by reducing the federal tax liability of the system's owner. To benefit from the tax credit, you have to have tax liability, which means you have to have sufficient wealth and income to pay taxes in excess of the value of the solar tax credit. The result is that only wealthy individuals and companies with sufficient tax liability can get the benefit; while nonprofits and people with less income pay more for solar. By prioritizing tax relief for the wealthy as a means to incentivize solar, we have created yet another system in which those with wealth can become owners while everyone else has to pay rent. Valuing wealth over well-being and money more than people got us here—but it won't get us where we want to go. We can do better!

Put the public *back in public utilities.* Whether your local utility is an investor-owned monopoly utility, governed by a public utility commission; a public power utility, owned and operated by a local government; or a rural cooperative, owned and operated by its members, all US utilities are intended to be governed for the public benefit. The notion that profitability, and the political and economic interests of faraway investors, is paramount clashes with the mandate of our public institutions. Moreover, this idea is fundamentally at odds with the democratic governance models of cooperative and public power utilities. Good governance that puts the *public* back in public utilities is an essential first step toward building clean energy futures that work for all.

Repair, restore, and do justice. Building clean energy futures presents an important opportunity to repair and restore communities and to do justice. Energy systems of the past that remain in operation hurt individuals and disproportionately harm Black, brown, and Indigenous communities. This damage is not by happenstance; it is by design. Racial segregation, discrimination, and other outgrowths of racist beliefs and white supremacist values were the law of the land when our current energy systems were being built. Among the consequences are that Black people in America are three times more likely to die of pollution than white people because polluting industries were more likely to be built in or near Black communities.[2] These wrongs must be righted during our lifetime. Restoring justice in our energy systems is part of the work.

Count your blessings and work from abundance. It's always helpful to begin with gratitude. Before we shifted to fossil fuels, we were lighting city streets by killing whales and burning oil made from their boiled flesh. Fossil fuels enabled progress, though at a high cost. Their time has passed, and we now know better, so we can give thanks for what is good, repent for what was bad, repair and restore what has been damaged, and move forward.

Notably, the energy system we built in the nineteenth century was based on scarcity—scarce fossil fuels, scarce capital, and the notion that power was only for the few who could afford it. Scarcity creates disparities, and disparities undermine justice. Clean energy, however, is based on what is available in great abundance: sun, wind, and land. Take a minute to appreciate the renewable energy resources you have right in your own community and use that gratitude as the foundation for your clean energy future.

Share and serve. Once you know what you have in abundance, you know what you have to share. Because sharing is a two-way street, you can also count on others to share what you need with you. Service is a companion to sharing that is often overlooked or looked down upon in our culture, as if serving others were a chore for those of low social or economic status, and being served an exclusive privilege of the well-to-do and well connected. Serving others, however, and especially those who need help is a privilege that you can and should choose to do with joy.

Courageous innovation. Change is hard. Most people don't like change, especially changes that they don't initiate. Leading people through a change, like the transformation of our energy system, takes courage, especially when it requires owning up to hard realities about the consequences of continuing to burn polluting sources of energy. It also takes innovation to be faithful to a vision and invent what you need to get there. *Courageous* and *innovation* are two words that I first heard paired by my dear friend David Mueller (an early leader in corporate sustainability) before his passing, and when linked they describe a mightily helpful quality for building better futures.

Partnership and connection. Putting the transformation of two of the biggest industries on the planet—energy and transportation—together to rebuild the vitality of our hometowns is no small task. Look for partners that share your values, as well as linkages between the projects and programs you build together that strengthen and multiply their impact, keeping in mind that broadband is the thread that ties it all together.

Guided by these principles, we shall embark on an exploration of rural power, its historical roots, and the thousands of local energy democracies that are its legacy. These rural electric cooperatives and small-town public power utilities are central to the "how" of our story. We'll then examine the national context in which these local nonprofit energy democracies operate to understand why we have many local energy futures, not just one monolithic energy future.

Next, we'll dive deeper to understand the governance and business models of rural cooperatives and small-town public power utilities. Knowing how these local utilities make decisions, as well as how they earn and spend money, is essential to the work of building clean energy futures that revitalize our hometowns.

We will then examine five components of a clean energy future you can put to work for your community, including strategies and examples for how to get started:

- *Energy efficiency.* We begin with energy efficiency, which is the most established and accessible starting point and has the benefit of immediately improving people's homes.

- *Solar power.* We then expand our perspective to include renewable energy, specifically solar power, which puts clean energy on a community scale.

- *Resilience.* Next, we address resilience, including energy storage and microgrids, which strengthen communities and are at the innovative edge of current technology.

- *Electric vehicles.* The electrification of transportation is the next wave of transformation. It's big business and we will see why partnerships are needed to localize its benefits.

- *Broadband.* Finally, we discuss broadband, which is woefully lacking in rural America but necessary for a smart, modern energy grid. By bringing together clean energy and broadband access, our rural and small-town utilities can meet two needs at once. But for this upgrade to happen, local clean energy leaders will need to become broadband advocates too.

We conclude with an exploration of how thousands of clean energy futures can work together in harmony, building shared capacity for what communities have in common while preserving local options to tailor solutions to local needs. As someone who is from a small town and prefers to live in a small town, I make a point of discussing how this clean energy–powered rural revitalization can enable people to stay in their hometowns and raise their families there too.

I hope you'll join me and the many people highlighted in this book who are investing their love, commitment, and expertise in building local clean energy futures. And I hope that this book will provide you with the practical inspiration, useful ideas, and expanded sense of community you will need to do it.

CHAPTER 1
The Roots of Rural Power

We are experiencing a transformation of our energy systems, the scale of which has not been seen in nearly one hundred years. It will take an estimated $2.5 trillion of public and private investment to build new solar and wind farms, expand energy storage resources, put tens of millions of new electric vehicles on the road, and make our homes and buildings more comfortable and efficient.[1] Moving toward clean, renewable sources of energy and away from fossil fuels creates the opportunity to localize our energy systems in the process.

It is a particularly pivotal moment for the rural communities and small towns that many of us call home. Not only do they have the ample land resources needed to generate all that clean power, but they're also home to many of the corporate and industrial leaders that are driving energy markets forward. Perhaps most important, however, rural and small-town communities are primarily served by local, nonprofit utilities that are ideally suited for building local clean energy futures that can help our hometowns thrive. When governed according to their democratic roots, those local utilities can be engines for economic prosperity guided by the communities they serve.

There are nearly three thousand such utilities across America serving more than 27 percent of all utility customers and more than half the landmass of the United States.[2] Because they were created to be locally owned and controlled as nonprofits with economic development as a part of their core mission, they can connect the value of clean energy to the needs and priorities of local communities in ways that investor-owned utilities can't because the investor-owned model is geared toward delivering profits to faraway shareholders instead of keeping value close to home.

These nonprofit utilities were created nearly a century ago to lead the last transformation of our energy system—the drive to electrify rural America and bring power to the farm. To understand how we can use the same institutions to build local clean energy futures today, we will begin by examining their past.

Energy Poverty

The story of the rural electrification movement of the 1930s will sound familiar because it is echoed by today's clean energy movement. From the beginning of electrification into the 1930s, the lack of access to energy was a source of grinding rural hardship and poverty. Nine out of ten rural homes had no electricity, and the private power industry had no interest in serving them. The country was too thinly populated, and corporations couldn't make a profit running transmission lines to small towns or farms.

Take a moment to imagine what it would have been like to live without electricity and recall if you can the stories from your own family's history. No power made for hard labor, including what was then referred to as "women's work." Washing clothes or running a bath meant boiling water, cooking a meal required a wood-burning stove, keeping warm meant families had to harvest and cut their own wood for fuel. After the sun went down, candlelight or kerosene lamps were the only means of illumination, which meant children had little light by which to study even if they were able to stay in school. There were few conveniences in this life, which was made harder by the aftermath of World War I, the economic devastation of the Great Depression, and the environmental devastation of the Dust Bowl. The gulf between the haves and have-nots was immense and growing wider.

The New Deal, led by Franklin Delano Roosevelt, transformed life in rural America. It included programs specifically designed to turn the lights on, protect and conserve the environment, and put people back to work. President Roosevelt's New Deal programs built or repaired 650,000 miles of roads, 125,000 buildings, 24,000 miles of water infrastructure; planted 220 million trees; and delivered electricity to nine out of ten rural households by 1945.[3] Three million men were employed just through the Civilian Conservation Corps, doing work that we would now call "green jobs." These opportunities, however, were not available to all Americans. The Civilian Conservation Corps did not employ women and was racially segregated; New Deal investments in segregated infrastructure, such as housing, fortified the era's Jim Crow laws.

From Private Luxury to Public Utility

Before the New Deal, electricity was a luxury available only to the few who could afford it. During the late nineteenth century, electrification was provided by small private power plants and delivered using direct current transmission over short distances. Enabled by technological changes, including the development of alternating current transmission lines that could deliver high-voltage power over much longer distances, the electric power sector began consolidating through mergers and acquisitions.

President Franklin Delano Roosevelt, architect of the New Deal. (Photo by Margaret Suckley, FDR Presidential Library & Museum, public domain; accessed via Wikimedia Commons: https://commons.wikimedia.org/w/index.php?curid=43594838.)

Consolidation enabled these early utilities to take advantage of available economies of scale to increase profitability. The resulting holding companies created complex pyramids of nested ownership, which were used by utility companies to build scale and dodge state-level regulation.[4] By the end of the 1920s, just ten holding companies controlled three-quarters of the US market while only about 3 percent of rural farming families had access to power.[5]

After his election in 1932 and in the years following the 1929 stock market crash, President Roosevelt took on the holding companies and took up the banner for public power. FDR's famous Portland speech of September 1932 laid out his electrification agenda and made the case for the governance of the power sector for the public good, highlighting many themes and values that are no less relevant today:

> As I see it, the object of Government is the welfare of the people. The liberty of people to carry on their business should not be abridged unless the larger interests of the many are concerned. When the interests of the many are concerned, the interests of the few must yield. It is the purpose of the Government to see not only that the legitimate interests of the few are protected but that the welfare and rights of the many are conserved. . . .
>
> When the great possessions that belong to all of us—that belong to the Nation—are at stake, we are not partisans, we are Americans. . . .
>
> . . . Electricity is no longer a luxury. It is a definite necessity. It lights our homes, our places of work and our streets. It turns the wheels of most of our transportation and our factories. In our homes it serves not only for light, but it can become the willing servant of the family in countless ways. It can relieve the drudgery of the housewife and lift the great burden off the shoulders of the hardworking farmer.

. . . A utility is in most cases a monopoly, and it is by no means possible, in every case, for Government to insure at all times by mere inspection, supervision and regulation that the public get a fair deal—in other words, to insure adequate service and reasonable rates.

I therefore lay down the following principle: That where a community—a city or county or a district is not satisfied with the service rendered or the rates charged by the private utility, it has the undeniable basic right, as one of its functions of Government, one of its functions of home rule, to set up, after a fair referendum to its voters has been had, its own governmentally owned and operated service. . . .

As an important part of this policy the natural hydro-electric power resources belonging to the people of the United States, or the several States, shall remain forever in their possession. To the people of this country I have but one answer on this subject. Judge me by the enemies I have made. Judge me by the selfish purposes of these utility leaders who have talked of radicalism while they were selling watered stock to the people and using our schools to deceive the coming generation.

My friends, my policy is as radical as American liberty. My policy is as radical as the Constitution of the United States.

I promise you this: Never shall the Federal Government part with its sovereignty or with its control over its power resources, while I am President of the United States.

A fundamental part of FDR's vision was that public power would serve as a measure for judging private utility rates and services. Specifically, as he noted in the same speech, "The very fact that a community can, by vote of the electorate, create a yardstick of its own, will, in most cases, guarantee good service and low rates to its population."

Electricity for All

True to the president's word, in 1933, the federal government took its first major action toward rural electrification with the passage of the Tennessee Valley Authority Act, which created what is still the nation's largest public power utility—the Tennessee Valley Authority (TVA).

Bringing together energy, environment, and economic development, TVA's mission remains unchanged: "To improve the navigability and to provide for the flood control of the Tennessee River; to provide for reforestation and the proper use of marginal lands in the Tennessee Valley; to provide for the agricultural and industrial development of said valley; to provide for the national defense by the creation of a corporation for the operation of Government properties at and near Muscle Shoals in the State of Alabama, and for other purposes."

Once established, the TVA quickly undertook an ambitious plan to purchase land along the Tennessee River and its tributaries, a watershed that encompasses more than forty-one thousand square miles, to build hydropower plants that would deliver low-cost electricity to the region while providing flood control. Low electricity rates were also central to TVA's economic development mission, including bringing power to the region's rural communities, which were among the most impoverished in the country.

Threatened by the TVA's affordable electricity rates, private utilities fought hard against it. The campaign was led by Wendell Willkie, who later became president of the Atlanta-based utility holding company Commonwealth & Southern Corporation (today's Southern Company). Willkie even ran an unsuccessful campaign for president against FDR in 1940 as the Republican Party nominee. Republicans of the time claimed that the TVA, like many New Deal–era initiatives, was "creeping socialism." Coal companies joined private utilities in their critique, lobbying their concerns that coal-fired electricity would not be able to compete with public hydropower.[6] Opponents even took to the courts in their attempts to undo the TVA and protect their profit margins. In *Ashwander v. Tennessee Valley Authority*, 297 U.S. 288 (1936), the preferred stockholders of the Alabama Power Company fought all the way to the Supreme Court, which ultimately upheld the constitutionality of federal government's creation of the TVA.

While the TVA survived and thrived, including staving off multiple campaigns for privatization, utility companies and their allies were able to successfully block the replication of its success in other parts of the country. Originally envisioned as the first of many public power utilities that would set a benchmark for measuring the affordability of electricity, the TVA still stands as a singular large-scale example of how effectively public power can marry energy, economic development, and environmental stewardship. Ninety years later, the TVA serves more than ten million people with power in seven states, including Tennessee and parts of Kentucky, North Carolina, Virginia, Mississippi, Alabama, and Georgia; manages hundreds of thousands of acres of reservoir lands and associated public recreation sites; and helps attract billions of dollars of job-creating investments per year into Tennessee Valley communities.

Rural Electrification

The creation of the Rural Electrification Administration (REA) by Executive Order 7037 followed the TVA Act in 1935. The REA, which was later incorporated into the US Department of Agriculture and became the Rural Utility Service, was established to support electrification in rural areas. When the Rural Electrification Act of 1936 provided federal loans to build electric utility systems to serve rural areas, loan applications began pouring in from farmer-based cooperatives that were cropping up across the states. In 1937, REA developed the Electric Cooperative

Here It Comes

RURAL ELECTRIFICATION ADMINISTRATION
U. S. DEPARTMENT OF AGRICULTURE

The Rural Electrification Administration turned the lights on for farms. (Credit: Lester Beall, artist [1903–1969], public domain; accessed via Wikimedia Commons: https://commons.wikimedia.org/wiki/File:Light_-_Rural_electrification_administration_-_Beall._LCCN2010646236.jpg.)

Corporation Act as model legislation that states could adopt to enable the formation and operation of these not-for-profit, consumer-owned electric cooperatives to lead the way. Many of the same federal financing authorities and the resulting network of rural electric cooperative utilities remain today.

To make the case to the American public, the federal government also launched a robust documentary filmmaking program as a broad-based mechanism for communication and engagement. Film, in fact, was the only medium available to reach a national audience. Newspapers lacked national circulation, radio lacked visuals, commercially broadcast television didn't exist until 1939, but in 1930, 65 percent of Americans went to the movies every week.[7]

The US Film Service was led by Pare Lorentz, an acclaimed documentary director who rose to prominence with *The Plow That Broke the Plains*, which recounted the history and causes of the Dust Bowl. Lorenz and the US Film Service were among the sponsors of a highly influential documentary titled *Power and the Land*. Directed by Joris Ivens, a Dutch filmmaker known for covering political and revolutionary subject matter,[8] this thirty-eight-minute film recounted how REA-led electrification was changing life for the better for rural families through the story of one Ohio farm. Released in 1940, just one year before the US Film Service was disbanded by Congress, *Power and the Land* walked the line between reporting and propaganda and helped propel the rural electrification movement forward. It not only portrayed the hardships of farm life without electric power but also lifted to heroic stature the linemen who brought electricity to rural America.

At the same time the federal government was focusing on rural power, it was also reining in the monopolistic abuses of private utilities. The Public Utility Holding Company Act of 1935 outlawed the pyramidal, nesting doll structures that utilities including Commonwealth Edison (which today is ComEd, owned by multiregional utility giant Exelon) had used to inflate stock values and utility rates and to avoid state regulation. That same year, the Federal Power Act gave the Federal Power Commission the authority to regulate the interstate transmission and sale of wholesale electricity and the responsibility to ensure that electricity rates were "reasonable, nondiscriminatory and just to the consumer."

At the local level, similar to rural cooperatives, public power utilities had begun forming as a grassroots response to the failures of private utilities. Like schools, parks, and libraries, public power utilities were created as units of local government to serve small towns and communities with no access to electric power. The very first public utility was born in March 1880 in Wabash, Indiana, when a group of local men hitched a threshing machine to the courthouse and turned the lights on. By the end of the 1920s, there were more than three thousand such public utilities, though their ranks shrank by more than a third over the following decade under pressure from private power providers. Changing technologies, and specifically the rise of diesel generators, made small-scale power efficient again, supporting a resurgence. By the end of the 1930s, the number of local public power utilities had grown to more than two thousand, about the same number that remain today.[9]

The structure of the US electric power sector that resulted from rural electrification, public power utilities, and the regulation of privately owned utilities for the public good delivered power to 90 percent of rural households within twelve years and kept utility rates and corporate growth steady for decades, until the onset of the energy crisis of the 1970s. President Carter made energy a central pillar of his agenda, and two pieces of legislation passed during his administration ended up setting the stage for the privatization and deregulation of the utility sector that would follow.

Deregulation and Restructuring

In the late 1970s, landmark legislation that was intended to increase competition and advance alternative energy, including renewables, had an unanticipated and more wide-ranging set of consequences. The Public Utility Regulation Policies Act introduced competition into the electricity generation market, which had previously been regarded as a natural monopoly and regulated as such. During the same period, the Department of Energy Organization Act created the US DOE and consolidated statutory and regulatory authority, budgets, and people from across more than a dozen federal agencies. The same legislation also replaced the Federal Power Commission with the Federal Energy Regulatory Commission, establishing the two primary national institutions with authority to shape the US energy sector.

While these Carter-era laws created the legal framework that enabled deregulation, the Reagan era elevated deregulation to a national policy priority as a part of a broader restructuring of the US economy. The deregulation agenda grew to include the energy, telecommunications, and aviation sectors, all of which had previously been viewed as requiring governmental oversight to protect the public good.

Over the next decades and into the 2000s, the advancement of deregulation in the energy sector opened some regional and state energy markets to clean energy by enabling competition and encouraging better climate outcomes with incentives that helped solar and wind compete. By delegating the responsibility of delivering the most economically efficient energy to the market, however, deregulation risked the original public purpose of federal involvement in the energy sector as articulated in the TVA Act in 1933: to deliver affordable power for all, to support economic development, and to preserve the environment.

Deregulation has been followed by the reconsolidation of the electric utility industry. Just since 1995, the number of publicly traded electric utility companies has declined by 57 percent. There are fewer than forty electric utility holding companies remaining, which together control more than 70 percent of US residential electricity access.[10] That's better than in the 1920s, when just ten holding companies controlled more than 75 percent of the market, but as consolidation continues, it's reasonable to wonder if the United States will return to the degree of concentrated control over energy markets that prompted FDR to act.

As a further outcome, some vertically integrated utilities are beginning to sell their generation assets in order to become "wires only" companies. Utilities own the transmission and distribution lines that are necessary to get electricity to market, and the high costs of building new transmission and distribution infrastructure keeps other players from entering the market. Because utilities are likely to retain control over delivering electricity and related services to customers, utilities that become wires only companies could profit from their monopolistic access to our homes and businesses through their electricity distribution systems—similar to the broadband companies that became successors to the old regional telephone monopolies after deregulation and the debut of the internet.

If the combination of deregulation and new technology in the energy sector does in fact follow the same pattern as the telephone industry, competition could lead to lots of innovation and new consumer services, but with rising consumer costs. This pattern is evident in what we pay for communications and connectivity today. The average US household pays $127.37 per month for cell phone service and $68.58 for internet access—nearly $200 per month combined—while the average electricity bill is $115 per month.[11] If electricity and related energy services are public goods that are fundamental to wealth and well-being, and the electric utility sector is evolving in the same direction as the telephone industry, what will be the right balance between universal access, competitive markets, and innovation to keep access to electricity affordable for everyone?

Rural Power

The questions we now face are similar to those we answered a hundred years ago by forming rural cooperative and small-town public power utilities were formed to deliver affordable power to everyone. This time, instead of rural electrification, we're cleaning up and modernizing the electricity sector, and the investments in clean energy technologies and infrastructure that are necessary to create new economic opportunities at a massive scale. This time, however, rural cooperatives and public power utilities are already in position as a force to drive the value of energy into local community development priorities.

Cooperative and public power utilities are part of the New Deal's long-lasting legacy. Including both rural electric cooperative and small-town municipal utilities, they are community-owned nonprofits that provide power to more than forty-nine million Americans. Together, they are nearly three thousand strong and serve all fifty states and all US territories.[12] While cooperative and public power utilities serve far fewer customers and have vastly less revenue than investor-owned utilities, their leadership and governance are much less concentrated and closer to the people, and their territory covers the majority of the land, which is part of their strength.

Public Power Utilities

There are 2,003 public power utilities serving more than twenty-two million customers; each is a unit of local government typically overseen by an elected city council or utility board. Of note, not all public utilities serve small towns and rural communities. The giants among them include the Los Angeles public utility, which is called the Los Angeles Department of Water and Power, the Puerto Rico Electric Power Authority, CPS Energy serving San Antonio, and the Salt River Project, which is an agency of the State of Arizona that serves the Phoenix area.

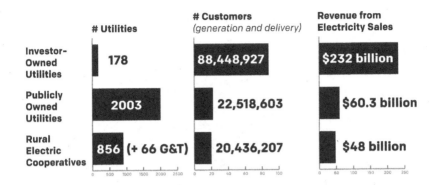

	# Utilities	# Customers (generation and delivery)	Revenue from Electricity Sales
Investor-Owned Utilities	178	88,448,927	$232 billion
Publicly Owned Utilities	2003	22,518,603	$60.3 billion
Rural Electric Cooperatives	856 (+ 66 G&T)	20,436,207	$48 billion

Types of utilities and their customers. (Source: 2021 Statistical Report, American Public Power Association, 10–13, https://www.publicpower.org/system/files/documents/2021 -Public-power-Statistical-Report.pdf; original graphic.)

Service Territories of Rural Electric Cooperatives

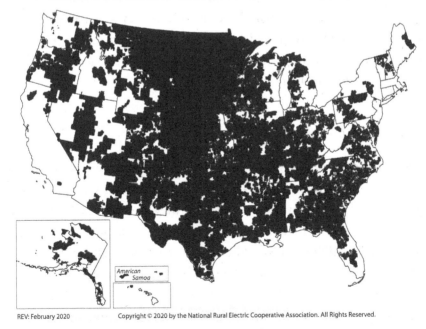

Service territories of rural electric cooperative utilities. (Original graphic.)

Locations Served by Public Power Utilities

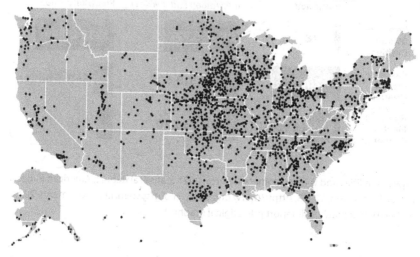

Locations served by public power utilities. (Original graphic.)

While some large public power utilities may struggle with maintaining governance that is proximate to the people they serve, all their missions are aligned with providing for the public good. Utility revenues are typically invested to expand or improve public services depending on each utility's public mandate.

Rural Electric Cooperatives

Rural electric cooperatives are member-owned nonprofit utilities, and there are 922 of them in the United States, including 66 that provide generation and transmission.[13] Together, they cover 56 percent of the land, own and maintain 2.6 million miles of distribution lines, and serve twenty million households, or 13.6 percent of all US electricity customers. Rural electric cooperatives were created to support local economic development, and their impact remains potent. As of 2017, they employed 68,200 people directly and an additional 268,500 contractors and suppliers, and they contributed more than $12 billion to local economies.[14] Cooperatives are owned and controlled by their customers, who are referred to as members, through democratically elected boards of directors. Of critical importance, rural cooperative utilities serve more than 92 percent of the persistent poverty counties in the United States, which have poverty rates ranging from 20 percent to more than 60 percent.[15] That makes rural electric cooperatives a notable source of community wealth and economic development, and essential agents to fight against rural poverty and for better quality of life.

Small Is Mighty

Today, as in the past, more than 70 percent of the electricity market is controlled by a small number of large utility holding companies owned by investors. Deregulation paired with clean energy incentives has increased competition and innovation, but at the risk of losing the public purpose of delivering affordable power to everyone. While investor-owned utilities pursue further consolidation and some sell off their generating assets to maximize profits for their shareholders, rural cooperative and small-town public power utilities can build clean energy futures a different way.

Created as a response to the failure of private industry to deliver power to small towns and farms, these local nonprofit utilities are the result of our energy system's last large-scale transformation. Purposely formed to connect energy to the economic well-being of the communities they serve, cooperative and public power utilities serve more than one in four US utility customers, and their service territory covers every state and US territory and more than half of the nation's lands. Individually, they are small, but together they are a mighty force that can show how to put the public good instead of profitability alone at the center of the public utility sector, thereby demonstrating how investor-owned utilities can do better. Moreover, because they are designed to be locally owned and controlled, they are ideally suited to connect the benefits of clean energy to local quality of life, enabled by the localization of our energy systems. By focusing on the power and potential of these nearly three thousand local nonprofit utilities, rural America can take the lead in building local clean energy futures that work for everyone.

As you will note, *futures* is plural. That's because the systems that provide, decide, and produce energy not only define the landscape in which local utilities operate but also empower us to define our own futures. Clean energy's infrastructure, our country's governance, and the natural environment that provides us with resources all work together to support bottom-up strategies, not top-down mandates. In the next chapter, we'll shift from the past to our futures and explore those strategies.

Being small and distributed instead of big and centralized doesn't limit the potential of rural and small-town utilities to serve as powerful agents of local economic development. In the past, the industrial scale necessary to extract, burn, and distribute fossil fuel energy was a constraint, but distributed renewable energy alleviates it as a factor. Sun and wind are everywhere, and rural and small-town utilities serve the vast majority of the land needed to site clean generation. In addition, distributed energy technologies are rightly deployed at the local scale. These developments, and the sustained presence of public power cooperative utilities in rural and small communities, mean that big power's hold on our energy futures can come to an end. We can shift from extraction that drives profits to restoration that serves people.

CHAPTER 2

Localization

Our Energy Futures

We've met the nearly three thousand local nonprofit utilities that built rural power in the 1930s and are ready to build clean energy futures today. Now we'll take a step back to look at the big picture and explore the landscape in which they operate. In this chapter, we examine three systems that shape our energy futures—infrastructure, governance, and nature—with a focus on how they work together to enable localization, the key to building a clean energy future for your hometown.

Natural systems that provide the resources we use to produce energy, and the governing systems we use to make decisions, have always been diverse, distributed, and optimized at the local level. Until the emergence of clean energy technologies, energy infrastructure was the odd man out. The industrialized power plants and long transmission lines introduced in the nineteenth century, along with the extractive processes that fueled them, operated at a scale beyond anything that could be designed for any one community's needs—moreover, they destroyed the places in which they were located. Distributed clean energy technologies, by contrast, bring our energy systems into alignment with local priorities and needs.

An alignment of local infrastructure, local governance, and local natural resources enables us to imagine our own clean energy futures instead of waiting for the future to happen to us. This system of interconnected systems has a degree of complexity, however, which is important to understand because it sets the boundaries, shapes the markets, and defines the opportunities that inform how we'll build the futures we envision.

As my friend Ari Wallach, a noted futurist, has observed, *futures* is not only plural but it's a verb, too, so next we'll review how each of these systems works to inform how we "future" our energy systems. As we look forward, as Ari would say, let's make sure the energy futures we build make us great ancestors one day.

Infrastructure

Let's start with infrastructure, because the localization of our energy systems thanks to clean energy is what makes our rural renaissance possible. If we were still working with fossil fuels that required drilling, burning, long power lines, and extractive economic models to keep the lights on, we wouldn't have much to talk about.

Before clean energy technologies emerged, energy infrastructure had to be big, industrial, and far away—connected with the homes and buildings that use energy through local distribution systems and long wires. By contrast, clean energy infrastructure can be distributed and may include local renewable power production on our rooftops and in our neighborhoods, complementing utility-scale production on farms and large tracts of land. Related technologies like energy storage can not only increase resilience but also improve the efficiency of our energy infrastructure and even lower its costs in the long term. As a result, our energy infrastructure can look more like healthy networks in nature—distributed, highly connected, and with production located much closer to end use. This transformation from polluting power plants to solar farms also helps to realize the potential of our thousands of cooperative and small-town public power utilities. Whereas before, they could only resell and distribute electricity, now they can own it and keep its economic value close to home.

OUR OLD ENERGY SYSTEM

Our old energy infrastructure: centralized, with big polluting power plants and long transmission and distribution lines. (Original graphic.)

OUR NEW CLEAN ENERGY SYSTEM

The new clean energy infrastructure: localized, distributed, renewable, and connected.

It's worth appreciating how revolutionary this is by looking at the present-day consequences of the energy infrastructure we built in the nineteenth century. Big mining, oil, and gas operations and big power plants—all connected over millions of miles of overhead wires—concentrated economic and political power in the hands of the very few who controlled the capital necessary to keep the system running. In the hundred-plus years since the first coal-fired power plants went into operation, the industrial energy system turned on the lights and powered a thriving economy that lifted millions out of poverty, but it also knowingly destroyed the health of many of its own workers, took the lives of people living in "sacrifice zones" near the worst of its pollution, and continues to drive the cycle of climate change that's threatening the civilization electricity is meant to sustain.

This industrial energy infrastructure was built on an extractive economic model that took natural capital mined with undervalued labor and burned it for the financial profit of a few faraway owners to produce electricity. This same model remains in operation today. Coal-extraction practices such as mountaintop mining—in which giant machines scrape the tops off mountains and push the rocks and dirt into waterways—destroy entire landscapes. Drilling for oil and gas pollutes wells that provide people with drinking water. The poorest families, who also tend to be disproportionately impacted by pollution, suffer the highest energy burdens, typically paying 14.5 percent of their entire household income to keep the lights on. The rural poor pay even more, up to 30 percent or more in many counties.

Communities located near refineries and power plants are plagued with disease. As NAACP has highlighted, the associated racial disparities are glaring. Sixty-eight percent of Black and African American individuals live within thirty miles of a coal plant and are twice as likely to die from asthma than white Americans. Only 1.1 percent of those employed in the energy industry are Black, while Black households comprise more than half of those paying 10 percent or more of their income to keep the lights on.[1] Moreover, Black and Latino households pay almost three times as much for energy as higher-income and white households.[2] It's clear that living near old energy infrastructure not only lacks benefits but could also cost you your life. It's not fair, it's not right, and it's not the way it has to be. Liberated from big polluting energy infrastructure, our energy systems can now work as they should by putting local communities and local futures first.

Governance

Energy policy plays a major role in shaping energy markets, and the policy environment is constantly changing and evolving. Every clean energy strategy you consider will need to begin with an assessment of how policy creates possibility, so understanding how federal, regional, state, utility, and local energy policy decisions come together is particularly important. If you run into a policy roadblock—say, building codes that are so old they don't include solar—this review will also give you a road map to going through or around it.

At a high level, governance is simply the mechanism by which communities of people make decisions, and the closer decisions are to the people they affect, the more informed the decision can be by experience as well as expertise. That said, while many important energy policy decisions are made by units of government at the local or state level, there's nothing simple about the overlapping jurisdictions and layers of government that shape our energy markets. The policy landscape is populated by an alphabet soup of agencies as diverse in purpose as the people and places they represent and serve. That's why our discussion here will focus on the principles that determine which level of government does what; the roles federal, state, and local governments play; and the function of some of the most important players.

Principles: Federalism and Pluralism

Two principles that are foundational to our constitutional democracy shape how we make energy policy decisions. These are federalism, which brings states together under an overarching federal government within a political system that also enables each state to maintain its own governing integrity; and pluralism, which disperses decision-making power across a diverse group of institutions instead of being concentrated in any one authority. Inherent in both federalist and pluralistic approaches to organizing decision-making systems is the idea that diversity is good and no one person or small group of people or interests should have singular authority over everyone else. To coexist, share power, and get things done, centers of decision-making have to communicate and compromise, and the integrity of the system as a whole depends on transparency and equal access. Combined, federalism and pluralism bring together people from different places who have disparate experiences and perspectives.

Taking a look at what happens when these principles are violated is perhaps the most helpful way to understand how important they are. For example, the oil and gas industry has recently targeted state legislatures with lobbying campaigns for legislation that prohibits local governments from implementing 100 percent renewable energy goals through building electrification, protecting the industry's economic interests. Identical legislation that preempts local authority by making it illegal for cities and towns to electrify buildings by eliminating new natural gas service has been introduced and is being passed in states including Arizona, Georgia, Louisiana, Oklahoma, and Tennessee.[3] By using its disproportionate access to power and money to shop for policy-making forums that are friendly to its political interests, the oil and gas industry is undermining the ability of local people to make local decisions. The results don't just burden communities with more unwelcomed pollution; they also distort how our federalist system is supposed to work.

Federal Roles

The federal government plays a number of important energy policy roles—among them, defining common goals, setting health and safety standards, enforcing compliance, investing in innovation and infrastructure, and regulating interstate commerce, which is especially important for enabling efficient markets, since every state has its own set of energy policies and rules. How to meet the resulting federal requirements is largely left up to states.

Federal roles in energy policy are distributed across a highly diverse community of agencies and commissions, some of which are responsible to the president and to Congress, and some of which are responsible to the courts. The people who staff those institutions are a combination of elected and appointed officials and career federal employees, so no one political party or set of interests ever has complete control. The dispersal of energy policy authority across the federal government is an example of pluralism at work. It may seem inefficient, but in government, sharing power across a system means that many potentially differing perspectives have to be considered before a decision can be made. As a practical matter, shared power is among the reasons we have energy democracies instead of an energy empire; it enables you to define local energy futures instead of being told what to do.

To get a better understanding of the federal role in energy and how it can support your vision, we'll focus on the executive branch and the federal agencies. We won't try to cover them all, or even to cover the most important agencies comprehensively. We'll simply introduce a few key players and explain what they do as guideposts to orient you.

Let's start with the US Department of Energy (DOE) because it has *energy* in its name, though the agency has different functions than you might assume. Viewed from the perspective of which responsibilities take up the largest percentage of its budget, DOE's primary role is overseeing and regulating nuclear power plants, including safety. DOE also encompasses an enormous research and development enterprise, including the national laboratories. The department's seventeen research institutions—the National Renewable Energy Laboratory in Colorado and Oak Ridge National Laboratory in Tennessee, to name just a couple—grew out of federally funded atomic research during World War II and are located all over the country. Owned by the federal government and operated by contractors, the national laboratories are responsible for many important discoveries and technological innovations that are helping to make our energy systems cleaner, more efficient, and more equitable. Scientific research led by the national labs also helps to inform other DOE responsibilities, such as setting energy efficiency standards for appliances like refrigerators and dishwashers, which helps protect people from poor-quality products that may look okay but would drive up the bills. Because of the scope of its authority and its focus on R&D, DOE can be an important source of technical assistance, research funding, and grants that support early-stage innovation.

Perhaps the most important federal agency for communities served by rural cooperative and small-town public power utilities, the US Department of Agriculture provides grant funding and low-cost financing for energy programs and infrastructure. Legacies of New Deal–era initiatives that helped electrify America, especially important programs include the Rural Economic Development Loan and Grant program, the Rural Energy Savings Program, and the Rural Energy for America Program. Together, these programs make billions of dollars of grants and very low-interest loans available through rural utilities.

The US Environmental Protection Agency (EPA) is responsible for protecting human health and the environment from pollution, including air pollution and other kinds of pollution from power plants. Through EPA's regulatory processes, which include many opportunities for members of the public to provide feedback, the agency sets national standards and schedules that power plants have to meet to stay in compliance. Air pollution isn't limited by state or even national boundaries, so national standards to regulate power plant pollution make common sense. EPA also advances environmental justice, which includes helping communities that suffer from especially high rates of diseases like cancer because they're located near industries that pollute. If part of your clean energy future is cleaning up after someone else's dirty energy past, EPA is one place to go for help.

The Federal Energy Regulatory Commission (FERC) is an independent federal agency that was created in 1977 to regulate the interstate transmission of electricity, gas, and oil, among other responsibilities related to interstate energy markets. FERC is considered "independent" because it is overseen not by the president but by Congress, and its decisions can be appealed before the federal courts. FERC's regulatory authority sets the "rules of the road" for how electricity, for example, is sold and transmitted across state lines. FERC very specifically does not have the authority to regulate the sale or transmission of electricity within a single state. That means the sale and transmission of electricity from your local utility to your home is regulated at the state or local level, depending on where you live. But we'll get to that a little later.

Other federal agencies with important roles include the US Department of Transportation, which helps set rules and standards for the transportation sector and whose work will help local governments electrify transportation; the Department of Defense, which is the largest energy consumer in the US economy; and the White House Council on Environmental Quality, which leads the implementation of the National Environmental Policy Act protecting the ability of US residents to be involved in major federal decisions that impact our environment.

As you can see from these few illustrations, while the federal government does play a number of critical regulatory roles related to energy that enable markets and protect the public, most of its work is to provide funding, financing, and other kinds of support, including programs specifically created to help rural cooperative and small-town public utilities. So, as you're envisioning your clean energy future, consider federal agencies as potential partners. We'll explore some specific examples of how they can help when we get into our later discussions of energy efficiency and solar power.

Regional Roles

Some aspects of our energy systems are governed at the regional level, which is necessary to create a shared set of rules that enable, for example, the transmission of electricity across state lines. In addition, some energy resources such as hydropower don't fit neatly into state geographic boundaries, so there are regional authorities that manage these shared resources.

Regional transmission organizations, or RTOs, are designated and regulated by FERC to oversee the operation of the electricity grid in a given region and are governed by independently appointed boards of directors. Independent system operators, or ISOs, have similar responsibilities and are also overseen by FERC; however, an ISO can cover just one state and has somewhat fewer responsibilities as it relates to the transmission network. Notably, not every state is part of an ISO or RTO. Altogether, this patchwork of interconnected wholesale energy markets and electricity transmission and distribution systems makes up the US electricity grid. These interconnected grids are incredibly important to our security, safety, and quality of life. When the Texas grid, which is called ERCOT, went down in February 2021, millions of Texans went without heat or power during freakish freezing weather linked to climate change, and many Texans died from the cold.

There are also a number of power marketing authorities that were created by the federal government to sell hydropower from dams that are operated by the US Army Corps of Engineers and the Bureau of Reclamation within the US Department of the Interior. The power marketing authorities make electricity generated by hydropower available at the lowest possible rates, primarily to rural and public power authorities, to support affordability. This is distinct from the role of the Tennessee Valley Authority, an independent federal power company whose mission includes economic development and environmental conservation.

State Roles

Each of the fifty states and US territories, as well as the District of Columbia, sets its own energy goals and regulates the energy market within its territorial boundaries, all within the rules set by the federal government and the regional energy markets that each state may (or may not) participate in. This is where we begin to see tremendous, even unexpected, variety in how Americans who may even be neighbors straddling a state line define their energy futures. State energy policy plays a much bigger role in determining what kinds of clean energy strategies work where you live than the federal government. Let's explore four examples to see how.

As of 2021, seventeen states and the District of Columbia have deregulated electricity markets. That means you don't have to be a utility to sell electricity, and residents and companies can choose among electricity suppliers. Deregulation of electricity markets means, for instance, that solar developers can build solar farms and sell the electricity they generate to other people or companies so long as the necessary companion laws and regulations are in place that enable all parties—the owner of the solar farm, the person or company buying the electricity, and the utility that owns all the wires in between—to keep count and settle up.

The roster of states that remain regulated, which means that it's illegal for anyone other than the utility to sell electricity unless it's the utility who's buying, is perhaps unexpected. Conservative or "red states," including Mississippi and Alabama, prohibit the free and open market sale of electricity, while more progressive or "blue states," including New

York and Illinois, enable anyone who can generate solar energy on their property to use as well as sell it to others within the state's geographic boundaries.

Many states and the District of Columbia have passed laws that set clear climate goals, including setting target dates by which the jurisdiction will achieve specific renewable and clean energy targets. These targets, typically called renewable portfolio standards (RPS), have been adopted by thirty-one states and territories to date, four of which have set 100 percent renewable energy goals. Setting an RPS can make solar, wind, and other renewable sources of electricity more valuable than fossil fuels that create pollution. Both the damage caused by pollution and the cleanup it requires have a downstream cost that isn't included in the upfront price of energy, though individuals and society as a whole have to pay for it eventually. So setting an RPS that makes clean sources of electricity more valuable is a practical way to incorporate future costs into today's energy prices, thereby creating a helpful market mechanism around shared clean energy goals. If you live in a state with an ambitious RPS or similar policy and you install solar panels on your property, your solar panels will pay for themselves faster because you get not only savings on your utility bill because you generated your own electricity, you get value from the associated renewable energy credits besides.

Rather that setting common climate and renewable energy goals that foster a statewide energy market, some states instead create more granular rules and regulations. For example, South Carolina has a combination of state-level solar tax incentives and utility-level solar incentives that works alongside the South Carolina Energy Freedom Act of 2019, which removed restrictive mandatory caps on how much rooftop solar could be installed in the state. Previously, state law limited the number of South Carolinians who could purchase and install solar panels on their own property. Georgia has taken a slightly different approach, one that Tim Echols, an elected member of the all-Republican Georgia State

Public Service Commission, has described as "Red State Solar." Georgia sets mandatory solar capacity and cost targets for the state's monopoly utility to meet. As a result of these mandates, the installed capacity of solar has grown steadily, though Georgia residents and communities remain very constrained in their freedom to own their own means of generating electricity, even on their own property.

Every state and territory, along with the District of Columbia, also has a public utility commission (PUC), public service commission (PSC), or similarly named entity that is responsible for overseeing the state's monopoly electricity and natural gas utilities, and sometimes other utility services. A fundamental purpose of these commissions is approving the rates that utilities can charge, thereby protecting consumers from potential price gouging by monopoly energy suppliers. This role is particularly important in regulated states that don't allow any competition, limiting or eliminating people's choices if they don't like the pricing or service they receive. Typically, PUCs don't regulate rural electric cooperatives or public power authorities, because these types of small utilities are governed by boards or municipal officials who are elected by residents and are therefore considered to be regulated at the local level. Members of these regulatory bodies can either be appointed or elected, depending on the state. Sometimes state PUCs suffer from "regulatory capture," which is when commission members prioritize the profitability and political interests of the investor-owned utility's stockholders over the interests of local residents who are captive customers. The way to defeat regulatory capture is to elect new people to serve on the PUC.

Utility Roles

States may have one or many utilities, each with its own service territory. For example, my home state of Georgia has one large investor-owned utility, fifty-two public power or municipal utilities, forty-one rural electric cooperative utilities (also called EMCs), and multiple generation and transmission utilities, plus the Tennessee Valley Authority, that provide power to small local distribution utilities. Where you live determines which utility generates and delivers power for your home. Understanding the administrative policies of your local utility informs how you implement your vision, and whether your local utility is leading the way, helping as a partner, or standing on the sidelines.

Different utilities have different administrative policies, processes, programs, and requirements that implement state goals and policies. For example, if you want to install solar on your property, in most places, you have to connect your panels to the grid through your electricity meter, which is owned by the utility. The state PUC sets the high-level rules for how this works, but your local utility creates and administers the application process and determines how long it takes and how much it costs to connect your new solar panels. This interconnection process can make it very straightforward, or very hard and expensive, to go solar. For example, Alabama charges customers who install solar a monthly fee of $5.41 per kilowatt of installed solar capacity, which actually increases people's bills by more than $27 per month. Referred to by many as the Alabama Solar Tax, this PSC-approved utility policy prevents people from installing solar on their homes because it artificially increases the cost by more than nine thousand dollars over the typical useful life of solar panels.[4] In these ways, utility administrative policies, while they don't have the force of law or regulation, can have similar impacts.

Local Government

Local government agencies are the closest to the people, and they play essential roles in energy policy. In jurisdictions that have a public power or municipal utility, the utility is owned and operated by the local government and its revenues pay for municipal services that support

residents. And even where the local municipality doesn't include a utility, municipal permitting and regulations (including planning, zoning,
and building codes) shape our energy futures. Whether your local government is the utility, or whether you just need its permission to build
your solar project, local governments have the most flexibility to change
the rules and influence the decisions of state, regional, and federal partners along the way.

Here are two simple examples that illustrate this point. What is a
solar panel from the perspective of your local zoning and building code?
Building codes have been around for a very long time, and often don't
include any notion of what a solar panel is, whether it is attached to a
building or installed on a parcel of land or as a canopy over a parking lot.
In the District of Columbia, building permitting processes treat rooftop
solar the same as a penthouse, requiring that solar panels comply by
the same basic set of rules, including height and setbacks. Many other
new and sometimes challenging questions arise for local governments to
answer as our energy technologies continue to advance. For example, in
New York City, the fire department created a set of fire safety standards
for outdoor energy storage systems.[5] These standards inform the feasibil-
ity, cost, and extent of energy storage deployment and access to energy
resilience for residents across New York City, as well as in other cities
that follow the example they set.

Nature

The infrastructure and governance systems that enabled localized energy
systems are nested within the natural environment, which is the source
of all the energy we use. The diversity of the natural environment at the
local level determines the right balance between using resources to produce energy versus conserving them to support other systems, like recreation, water, or food. Whether we're digging up coal from under the

ground or harvesting the energy of the sun, the energy cycle is deeply connected to the natural environment both in terms of resources and impacts.

Our natural environment has an abundance of diversity, and that includes which clean energy resources are available locally. In some places it may be wind, in other places it may be land and solar—and some areas even have enough resources to share. Whatever the case, an inventory of what the local environment has to offer as well as the resources you need to protect is critically important because if your energy future doesn't work in harmony with nature, it won't be sustainable and therefore won't be much of a future.

To appreciate the diversity of the natural environment in the United States, we can look to the nearly one thousand Level IV ecoregions categorized by the EPA, each of which identifies an area where local ecosystems are similar. These designations provide a useful framework for scientific study and for managing natural resources,[6] but for our purposes, their abundant diversity illustrates another important facet of why, even when we embrace clean energy and a healthy climate as common goals, we don't have a "one size fits all" energy future.

The concept of ecosystem services is another helpful way to understand how the local attributes of the natural environment inform your energy future. Ecosystem services are the valuable services nature provides for people and society, although we don't typically assign a monetary value to these benefits unless we're extracting or harvesting them. Ecosystem services include provisioning services like drinking water and timber; regulating services like pollination and flood control; cultural services like inspiration and recreation; and supporting services like photosynthesis and the water cycle.

Energy systems can be analyzed based on their impacts on ecosystem services, providing a holistic perspective on how different clean energy technologies may or may not work in harmony with the places they're located or the areas they serve. For example, a windy mountain ridge might seem like the perfect place for wind turbines if you consider only how hard and consistently the wind blows, but some mountain ridges may be a poor option if they're in the path of migratory birds that could be killed by big turbines. Ground-mounted solar farms might be a good match for flat and sunny land, but some land may be more important to preserve for agriculture because of the qualities of its soil. The considerations can be complex, but it's worth taking the time to understand the full impact of our choices because our goal is using clean energy to revitalize the places we live, and we won't achieve that goal if we don't.

Agrivoltaics are a great example of how studying the interactions between energy systems and the natural environment can lead to better outcomes than looking only at energy systems. Greg Barron-Gafford, PhD, is a researcher at the University of Arizona. He works at the nexus of food, energy, and water, and has connected researchers and practitioners from all over the world to examine how co-locating solar installations and agriculture can benefit both. After planting food crops under slightly elevated solar panels in a hot, dry climate, Greg's research team documented increased crop yields, decreased evaporation and therefore irrigation requirements, and higher energy production because the area around the solar panels stayed a little cooler. Pairing energy systems and ecosystems in a thoughtful way leads to better results than focusing on either system alone.[7]

Empowerment

Now we understand how the larger energy landscape in which our local utilities operate empowers local energy visions and decisions. Clean energy technology opens up worlds of opportunity that weren't previously available because it enables localization. Leaving behind the big, dirty infrastructure of the nineteenth century also gives us the chance to heal the harms that its pollution has done to many of our neighbors, enabled by our governing systems. Although the number, diversity, and overlapping authority of the policy making institutions where we come together to make decisions about energy are complicated, the government agencies involved in energy can be viewed as helpers—setting goals, creating standards, offering financing infrastructure, and keeping us safe. Sustaining the entire system, nature provides the resources we need to realize our visions, so long as we use them in ways that don't harm or deplete other resources that we need.

Each of these systems comes together and connects to the lives of local communities through rural cooperative and small-town public power utilities, because the production, sale, and distribution of energy flow through them. Their missions, which include delivering power and economic development, connect the economic value and other benefits of energy to the communities they serve in ways that can either enrich or impoverish them, which underscores the leading role these nonprofit utilities play building our energy futures. Each utility's individual governance determines which of these paths it follows, so local utility governance is where we'll turn next.

CHAPTER 3

Thousands of Local Energy Democracies

Before you can get involved with your local utility to build a clean energy future for your hometown, you've got to understand how the utility makes decisions. Rural cooperatives and public power utilities were established as nonprofits that connect energy with economic development and as energy democracies. Whether managed by a board of directors or a city council, the governing bodies of both types of utilities are democratically elected by the people they serve. Democratic governance that gives customers and residents control is an important part of why local nonprofit utilities are positioned to be heroes in a rural renaissance, so long as they're governed well and with integrity.

As you'll recall, there are nearly three thousand such utilities, and while they serve just over 27 percent of US households, they cover more than half the nation's land and more than 90 percent of counties experiencing persistent poverty, which means their geographic service territories align with rural and small-town needs. By contrast, big investor-owned utilities produce and sell more electricity to more people, but from a rural perspective, they are less relevant because they primarily serve urban and suburban markets—a long-term legacy of their early refusal to run power lines to farming communities and small towns.

While the democratic governance of rural cooperatives and small-town public power utilities has long held extraordinary promise, that promise has never been fully realized. Since the early 2000s, research and reporting on cooperatives have documented failures ranging from financial abuses to manipulated election processes that prevent leadership change to decades of board supermajorities that don't represent the diversity of the communities they serve. To understand the difference between how these energy democracies were meant to be governed versus how many are falling short, we'll review their governing principles and share examples of governance gone wrong. We'll also explore what can be done to restore democratic governance by exploring two case studies that illustrate the importance of first principles and fresh leadership on the pathway toward a clean energy–powered revival.

Governance Structures

The governance of rural cooperative and public power utilities is distinguished by local control, democratic participation, and public accountability. The two models, however, have distinct governing structures, both of which operate in sharp contrast to the corporate governance of investor-owned utilities.

The roots of cooperative governance date back to 1844, when twenty-eight English cotton mill workers joined together as the Rochdale Society of Equitable Pioneers. Like many working people at the time, they faced low wages and miserable working conditions, and they couldn't afford household necessities, including food. The Rochdale workers determined that if they pooled their modest resources together, they could purchase goods at a lower price. The success of their endeavor—which initially sold only flour, sugar, oatmeal, and butter—led to greater ambitions; namely, a new type of enterprise that enabled members to share in profits and participate in decisions about the business. To guide this cooperative enterprise, they created the Rochdale Principles, which are still present today in the first four of the Seven Cooperative Principles.[1]

The International Cooperative Alliance (ICA), which was founded in 1895 and represents approximately three million cooperatives worldwide, maintains and publishes the Seven Cooperative Principles. According to the ICA, cooperatives are "based on the values of self-help, self-responsibility, democracy, equality, equity, and solidarity." Moreover, "in the tradition of their founders, cooperative members believe in the ethical values of honesty, openness, social responsibility, and caring for others."[2] The Seven Cooperative Principles, which are shared by rural electric cooperative utilities and embedded in the US regulatory definition of their nonprofit status, put these values into practice. These principles include open and voluntary membership; democratic member control; members' economic participation; autonomy and independence; education, training, and information; cooperation among cooperatives; and concern for community.[3] The governing structure of electric cooperatives proceeds from these principles.

There are two types of cooperative utilities: distribution utilities, which deliver electricity to residential and commercial members as centralized cooperatives; and generation and transmission utilities, federated cooperatives that buy and sell wholesale electricity for their membership, which consists of distribution cooperatives. Each cooperative utility has an elected board of directors, which hires and oversees the cooperative's chief executive officer, who in turns oversees and directs the cooperative's staff. Cooperatives are built on the principle of "one member, one vote," so it doesn't matter how big or how small a member may be as measured by their electricity usage—everyone is supposed to have the same voice.

THE SEVEN COOPERATIVE PRINCIPLES

1. Open and Voluntary Membership
Membership in a cooperative is open to all people who can reasonably use its services and stand willing to accept the responsibilities of membership, regardless of race, religion, gender, or economic circumstances.

2. Democratic Member Control
Cooperatives are democratic organizations controlled by their members, who actively participate in setting policies and making decisions. Representatives (directors/trustees) are elected among the membership and are accountable to them. In primary cooperatives, members have equal voting rights (one member, one vote); cooperatives at other levels are organized in a democratic manner.

3. Members' Economic Participation
Members contribute equitably to, and democratically control, the capital of their cooperative. At least part of that capital remains the common property of the cooperative. Members allocate surpluses for any or all of the following purposes: developing the cooperative; setting up reserves; benefiting members in proportion to their transactions with the cooperative; and supporting other activities approved by the membership.

4. Autonomy and Independence
Cooperatives are autonomous, self-help organizations controlled by their members. If they enter into agreements with other organizations, including governments, or raise capital from external sources, they do so on terms that ensure democratic control as well as their unique identity.

5. Education, Training, and Information
Education and training for members, elected representatives (directors/trustees), CEOs, and employees help them effectively contribute to the development of their cooperatives. Communications about the nature and benefits of cooperatives, particularly with the general public and opinion leaders, help boost cooperative understanding.

6. Cooperation Among Cooperatives
By working together through local, national, regional and international structures, cooperatives improve services, bolster local economies, and deal more effectively with social and community needs.

7. Concern for Community
Cooperatives work for the sustainable development of their communities through policies supported by the membership.

Cooperatives around the world are organized around the same seven core principles. (Source: National Rural Electric Cooperative Association; original graphic.)

Of note, democratic control is among the basic operating requirements that must be maintained in order for a cooperative to qualify as tax-exempt under the US tax code. Cooperatives that violate their democratic governance principles risk their nonprofit status.[4] How and when individual cooperatives hold board elections is described in their corporate bylaws, which are among the legal requirements for operating in most states.

Public power utilities, including municipal utilities, are similar to cooperatives in that the local community served by the utility has ultimate control of the utility's direction through the ballot box, but the model is different. Publicly owned utilities are a unit of local government, typically overseen by the city council or an elected board of directors. The city council or other elected governing board sets policy, including approving utility rates and programs, while the mayor hires and oversees the city manager and director of utilities. The Seven Cooperative Principles are not applicable, and governing values may shift with the election cycle, but ultimate control lies with voters, and the utility itself is publicly owned.

Some public power utilities, such as the City of Ellensburg in Washington State, take the additional step of creating a citizens' advisory council, which is a utility-specific forum for residents to engage and express their views. Since municipal governments that operate utilities and the city councils that oversee them have diverse responsibilities, particularly in larger communities, forming an advisory council can be an important way to bring more perspectives to the table to build consensus and thoroughly inform utility decisions.

By contrast, investor-owned utilities are for-profit corporations that operate to create income for their shareholders. Shareholders, also called stockholders, are investors who own equity in the corporation. According to data from the Federal Reserve's Survey of Consumer Finances, only about 53 percent of Americans own any stocks, and most are indirectly held through retirement savings or mutual funds. Stock ownership is inequitable. The top 10 percent of income earners own 10 times as much stock as the bottom 60 percent.[5] That means that for about half of Americans, ownership of investor-owned utilities is inaccessible and direct ownership and control are even further away.

The benefits of cooperative and public power utility governance are clear: ownership and control by local customers and residents versus faraway investors; access and participation by everyone versus just those who can afford it; and a mission to benefit people versus extract profits. These attributes enable cooperative and public power utilities to make decisions about energy within the context of the communities they serve, placing people at the center of their operations rather than exclusively seeking profit. The democratic cooperative governance model and governing structures of public power utilities should be ideally suited for building equitable clean energy futures, but in too many places, something has gone wrong.

Good Governance Gone Bad

Over the past two decades, numerous reports have detailed alarming declines in democratic governance among cooperative utilities, including extremely low voting participation and multiple documented cases of graft and abuse. America's energy democracies are filled with promise, but they need a grassroots reformation to reground them in the principles of the cooperative movement.

While democracy is closely held as a value, voting participation has been poor. When too few people with too little diversity control election outcomes, democracy breaks down. According to the Institute for Local Self-Reliance, 72 percent of electric cooperatives have less than 10 percent voting turnout on average.[6] A recent study from Portland State University showed the statistics for municipal elections are similarly dismal. Fewer than 15 percent of eligible voters participated in mayoral and city council elections, the average age of participating voters was fifty-seven, and there were wide disparities from neighborhood to neighborhood.

Low voter turnout can be due to lack of access, lack of awareness, or voter suppression, and all of the above have been on display. Self-dealing board nomination processes, failure to hold board elections, and corrupt use of proxy voting have all been documented through initiatives such as the Co-op Democracy Project. For example, in 2009, Randy Wilson famously ran for the board of the Jackson Energy Cooperative in Appalachian Kentucky—the first time any board candidate had ever been opposed in the cooperative's seventy-one years. After collecting five hundred signatures just to get on the ballot, Wilson lost the election by nearly six hundred votes. Only 2 percent of members cast a ballot, and most voted by proxy.[7]

Without broad participation and member engagement, it's easier for corruption to take hold. Tri-County Electric Cooperative in South Carolina, a small rural utility with some of the highest utility rates in the state, was revealed by *The State* newspaper to be paying each of its nine board members $52,000 per year—more than double the average per capita income of many of the communities Tri-County served—plus perks that included health insurance for life.[8] Just three months later, in August 2018, more than 1,500 Tri-County member-owners came together and voted the whole board out.[9]

It's easy to count the money and focus on the bottom line but much harder to measure the public good. The tendency of nonprofit utilities to lose focus with regard to their missions and behave like for-profit businesses can be observed in municipal utilities too. In "enterprise cities," revenue from the municipal utility is used to pay for essential municipal services, including community development, parks, and public safety, with all aspects of municipal government operating as a single enterprise. While this aspect of public power utilities can be beneficial and is critical to their ability to fund other community priorities with energy revenues, losing sight of the distinct public functions of individual units of government can become all to easy.

In my own hometown of LaGrange, Georgia, which is an enterprise city, the utility used to have a policy that required residents to pay nonutility debts to the city, including court fines, before they could get utility service. The impact of treating residents as customers of a single business instead of as citizens with a diversity of relationships with local government was discriminatory, and the city was sued in 2017 by a number of groups—including seven local residents who had been impacted and the Troup County Chapter of the NAACP. The local city council subsequently changed the policy and settled the lawsuit.[10]

Racial Equity

In addition, and of particular note, many cooperative boards do not reflect the diversity of the communities they serve. A century ago, at the time of their founding, racist Jim Crow laws across the United States institutionalized white supremacist systems and disenfranchised Black Americans. While these laws were dismantled in a series of hard-won Supreme Court victories culminating in the Civil Rights Act of 1964 and the Voting Rights Act of 1965, many of the economic, social, civic, and other structures they formed remained in place. So when we see a cooperative that has not had a contested board election in more than seventy years, we know that cooperative's governing practices were set in place during the Jim Crow era. These legacy governance systems, just like redlining practices in the banking sector that kept Black families from accessing financing for businesses and homes, perpetuate racial inequities from generation to generation until they're rooted out.

Racial disparities in governance access and inclusion among cooperatives are epidemic. For example, the Rural Power Project studied twelve electric cooperatives in the South for its *Democracy Lost and Discrimination Found* report. Across 313 cooperative boards with three thousand members, only 4 percent of board members were Black, whereas Black people made up 22 percent of the population. The picture was similar for Hispanic people, who composed about 10 percent of the population but only 0.3 percent of board seats, and for women, who represent a little more than 50 percent of population but held only 10 percent of board seats. The homogeneity of cooperative boards consisting of more than 90 percent white men corresponds with disparities in energy burdens, which are disproportionately carried by Black households, who represent 51 percent of all households paying 10 percent of more of their entire income for the electricity bill.[11]

Reform

Rural cooperatives, like publicly owned utilities, were built for local control. Reforming their governance systems can begin locally too. Local community development organizations like Kentuckians for the Commonwealth illustrate how not only by taking independent action but by coming together with similar groups to share resources, tool kits, and best practices to equip others to follow their lead.

The Eastern Kentucky Coalfield in Appalachia typifies the extractive economic model of the energy systems we built during the nineteenth century. The Appalachian region is rich in resources, but its people are poor, particularly in Appalachian Kentucky, where the poverty rate remains higher than in any other Appalachian region, exceeding 26 percent for more than forty years.[12] Coal mining took off there in the 1820s to feed the energy appetites of urbanizing nearby cities like Louisville, but while the coal mines provided work, they didn't build local wealth, because the economic value of Kentucky's coal was extracted by mining companies and exported to other regions.[13]

In 1981, Kentuckians for the Commonwealth (KTFC) was founded by a group of local citizens representing twelve Kentucky counties to call for tax reform, challenging the coal and timber companies that had been extracting wealth from the region for 150 years to pay their fair share.[14] Twenty years later, KTFC launched the New Power campaign for clean energy and a just transition focusing on eastern Kentucky, which not only included the state's most impoverished coal counties but its most energy burdened ones as well.

The New Power campaign focused on sixteen rural electric cooperatives that were part of the East Kentucky Power Corporation. As a part of the campaign, KFTC worked with local member-owners (including people like Randy Wilson who wanted to serve on their cooperative's board), and helped them unearth and decipher the relevant bylaws that defined board elections processes.[15] Obscure bylaws and cooperative bureaucracies helped to maintain the status quo. As Nikita Perumal of KFTC described on *The Next System* podcast, it was common to find cooperative board members that had held their seats for more than thirty years—before passing them on to their sons.[16]

While the New Power campaign struggled to get anyone elected to their local cooperative board, it revealed the scope of re-democratization reforms that were needed, engaged and mobilized local leaders, and led to the development of the 2012 Rural Electric Coop Reform Platform, which called for a restoration of cooperative principles through measures including mail-in voting, eliminating proxy voting, and member access to all board meetings.[17]

KTFC came together with organizations including the Mountain Association for Community Economic Development serving eastern Kentucky and southwestern Virginia and the Northern Plains Resource Council, which serves Montana, to create the Rural Electric Cooperative Toolkit. The tool kit includes information about board elections and bylaws, research reports, and campaign planning resources that can help local leaders assess the governance health of their own local cooperative and create a re-democratization plan if needed.[18] But for reform to lead to revival, it takes leadership from the inside too.

Revival

Curtis Wynn is no stranger to breaking barriers. As former CEO of Roanoke Electric Cooperative, he steadily advanced some of the most pioneering clean energy programs in the country. Now CEO of SECO Energy in Central Florida, one of the largest electric cooperative utilities in the country, he's far from done. In 2019, Wynn took the helm of the National Rural Electric Cooperative Association (NRECA), a national membership organization of America's more than nine hundred cooperative utilities, as its elected president for what would become one of the most consequential two-year terms for the future of cooperative governance in the organization's history.

Born and raised in Graceville, Wynn went to work washing trucks at the local West Florida Electric Cooperative while he was still in high school. In 1997, he became the first-ever Black CEO of an electric

Curtis Wynn, CEO of SECO Energy (Sumter Electric Cooperative). (Photo: Dan Charles / NPR)

cooperative. He viewed the transformation of the electricity business as an opportunity for putting cooperative member-owners at the center of innovation, and that's exactly what he did.[19] Roanoke Electric Cooperative's territory in northeast North Carolina, where Wynn was CEO, has an aging population, little growth, and poverty rates that are more than double the national average. Thanks to Wynn and his team, the people they served also have market-leading efficiency programs that help reduce bills, a community solar program that's building multigenerational wealth for local Black landowners, and a brand-new electric vehicle program that includes discounted subscription charging rates.

Wynn carried this same vision into his role as NRECA president. As he shared in March 2019 in Orlando, "During my tenure as president, I will be challenging all of us to fully examine the way we view our future, our individual cooperative's place in it, and how we can lead our co-ops and our communities into the next generation. . . . Our member-consumers are demanding more and more from us—whether it is more convenience, diversity of thought and perspectives, or evolving services. My simple suggestion is that we acknowledge that change is happening, act on it, and lead through it."[20]

Then came 2020. Rocked by a global pandemic and a racial reckoning ignited by the murder of George Floyd by a police officer in Minneapolis, the first Black president in NRECA's seventy-seven-year history would take on a new mission that would become part of his legacy: placing diversity, equity, and inclusion (DEI) at the top of NRECA's agenda through a new DEI policy that would be adopted membership-wide.

Making DEI the center of his platform aligned with Wynn's initial focus on transforming the electric utility sector without leaving anyone behind. Importantly, the timing coincided with NRECA's annual resolutions process through which new policies can be adopted across the whole of the cooperative utility community. As Wynn explained,

"We needed the membership to speak directly to DEI to give lasting direction."

Member resolutions not only give lasting policy-level direction to NRECA's CEO and staff, but the process itself educates the membership. Each year, new resolutions begin their journey in June with the NRECA's Resolutions Committee, are referred out to the regions who vote on them in the fall, and then return to the Resolutions Committee for final vetting in the winter before going before NRECA's more than nine hundred members in the spring for a full membership vote. As Wynn recounted, "The DEI resolution was received with a sense of hunger. People wanted to know, 'What is this?'"

Wynn went on the virtual road to support the resolution. While the COVID-19 pandemic precluded travel, he attended statewide and regional meetings and won an unprecedented level of support across the organization to move the policy forward. He explained to his fellow cooperative leaders how DEI is aligned with the Seven Cooperative Principles, which include open and voluntary membership and democratic control. He also made the case that DEI is good for business, not only because all cooperatives have a responsibility to engage all members, but because decades of business research show that diverse organizations also get better results.

In March 2021, as Wynn handed the presidency over to Chris Christensen of Montana, the DEI policy had passed with the unanimous consent of NRECA's more than nine hundred member cooperatives and the unanimous support of Wynn's fellow members of the NRECA's national board.

While Wynn was leading America's rural electric cooperatives through the DEI member resolution, he was also leading by example as the then CEO of Roanoke Electric Cooperative, where he had engaged a DEI consultant to help put specific strategies, goals, and measures in place to chart their direction and process. Roanoke Electric Cooperative had long led the nation in diversity since the majority-Black membership it serves organized in the 1970s and first won representation on the board.[21] The long-term results of board leadership that reflects the community it represents were evident in Roanoke's diverse workforce and in the range of innovative services it offers member-owners. Yet Roanoke didn't have a playbook it could share with other cooperatives that wanted to follow suit.

Documenting how Roanoke put its values into action would enable Wynn to help other rural cooperative leaders understand how to put their own DEI policies into practice too. "The neat thing about this is that there are so many people out here working on the same things," Wynn shared. "It's time for us to get in sync to move the needle." The third and final piece of Wynn's plan is Project Deliver, which puts DEI into an actionable framework that other cooperatives can use. Bringing together strategies that help cooperatives respond to the transformation of the utility sector with approaches that prioritize equity as a part of a broader DEI strategy, Project Deliver is a turnkey solution for others who want to follow Wynn's lead.

"It is a very pivotal moment," Wynn reflected. "The energy transformation, the focus on DEI and sustainability, and the scale of Federal investment in infrastructure—we have such a window of opportunity right now in terms of the way things could look on the other side. As a forty-year veteran [of electric cooperatives], I see the opportunity to use what the Lord has given me to lead in this clear direction."

External Pressure

Not all the challenges to democratic governance come from within local utilities. The ability of both cooperatives and small public power utilities to define their own futures, even when local governance is good, can be constrained by upstream relationships with generation and transmission utilities. Generation and transmission utilities were formed by local utilities so that they could pool their buying power, increase market leverage, and reduce costs. Local cooperatives created federated cooperatives, and small publicly owned utilities formed statewide associations. This structure enabled the local utility to focus on serving customers by distributing electricity while the generation and transmission utility engaged in generating and purchasing electricity at scale.

In these arrangements, local utilities purchase electricity at wholesale prices from the generation and transmission utility, typically on long-term contracts lasting at least twenty years. Where these long-term contracts have locked local utilities into purchasing coal-fired power and other polluting sources from fossil fuels, they have limited the ability of local utilities to switch to clean energy sources in response to member and customer demands. While the governing boards of generation and transmission utilities are made up of representatives of the smaller utilities they serve, in practice, the power dynamics can favor generation and transmission utilities at the expense of their members. For example, local utilities could decide they wanted clean power, but their contracts with upstream generation and transmission utilities could prevent them from implementing their decisions.

Futures-Focused

When Curtis Wynn speaks about cooperative utilities and the communities they serve, he speaks with love—love for the model, love for the people, and love for what cooperative utilities can do. He emphasizes the imperative of whole community engagement and the power of people-centered energy innovation. And he's right. Without good governance, utilities designed as local energy democracies just replicate the disparities of the nineteenth-century economy at a smaller scale.

Their fundamental governance design, however, also fills them with promise. Because their decision-making leadership is elected by members of the same local communities they serve, both rural cooperatives and small-town public power utilities possess the basic governing mechanisms that are necessary to place the economic benefits of energy within and in service to a community rather than the other way around. While many of these utilities have strayed off course, the road map for getting back on track is present in the power of engaging in local elections, and in the Seven Cooperative Principles themselves. Organizations like Kentuckians for the Commonwealth and leaders like Curtis Wynn are blazing a trail and creating tools many other communities need to re-democratize their local utility. Viewed from a national perspective, they demonstrate the importance of grassroots campaigns and courageous leadership at the top to enable communities to build equitable clean energy futures, everywhere.

Now that we have a picture of how local nonprofit utilities make decisions, we'll turn to how they earn and spend money. Understanding the business models of rural cooperative and small-town public power utilities is fundamental to knowing how to put them to work.

CHAPTER 4
Growth for the Greater Good

Now that we've studied the histories and governance of America's cooperative and public power utilities and the energy landscape in which they operate, it's time to dig into their business models. Understanding how nonprofit utilities earn money, spend money, and accesses capital to grow will show us how to connect the value of clean energy services like energy efficiency, renewables, resilience, and electric vehicles to the needs and priorities of the communities they serve. Access to capital to build local clean energy futures is essential, so we'll focus on the funding and financing options that are uniquely available to local nonprofit utilities.

Both rural cooperative and public power utilities have access to hundreds of billions of dollars of low-cost financing, zero-interest loans, and grant funding from government and nonprofit sources—all of which will only increase as federal and state governments invest in modernizing our nation's infrastructure. Because local nonprofit utilities are not dependent on big banks or other for-profit sources of capital that demand the lion's share of the profits as interest in return for lending money or as dividends for investing equity, the economic benefits can remain with member-owners and local residents who own and control the utility instead of flowing out of the community to whomever provided the capital.

While the scope and diversity of financing and funding options are expansive, the opportunities they create can be complex and hard to navigate. In this chapter, after we've described cooperative and public power utility business models and highlighted how they differ from investor-owned utilities, we'll survey and assess a number of their financing and funding tools in detail. Understanding the sources of capital that are specifically available to small nonprofit utilities will enable to you to develop the project and program finance strategies necessary to implement your vision.

Business Models

As with their governance structures, the business models of rural cooperatives and small-town public power utilities were formed to enable communities of farmers, workers, small businesses, and small towns to own and operate their own local utilities. They are distinguished from large investor-owned utilities (IOUs) by the fact that they are controlled locally and the profits they produce stay at home. Expanding the programs, services, and revenue of locally owned and controlled utilities creates local jobs, wealth, and value. By contrast, IOUs are privately owned monopolies that support local jobs but redistribute profits, value, and wealth from local communities to distant investors, though state-level regulation by public utility commissions is intended to strike a balance between the monopoly utility's profits and the public good.

Rural Electric Cooperatives

Rural electric cooperatives are nonprofit organizations as classified under section 501(c)(12) of the Internal Revenue Code. That means that, so long as they meet the requirements of the tax code, they do not pay federal taxes. This tax treatment dates back to the Revenue Act of 1916, which recognized organizations including "mutual ditch or irrigation companies, mutual or cooperative telephone companies, or like organizations" as tax exempt. Rural electric cooperatives were considered to be "like organizations," and this status was formalized by Congress in the Miscellaneous Revenue Act of 1980. To qualify for tax-exempt status under the amended code, the Internal Revenue Service requires an organization to meet three principles: be organized and operated under cooperative principles, adhere to the activities for which it was created, and derive no less than 85 percent of its income from members.

The first requirement, to be organized under cooperative principles, ties back to the Seven Cooperative Principles. Three activities are necessary to meet this test: the cooperative must be democratically controlled, operated at cost, and must put the interests of members ahead of any financial institutions that provide capital. Some of the ways that cooperatives meet these tests include holding regular membership meetings, democratically electing board members, distributing profits (which cooperatives refer to as savings) to members according to how much business individual members did with the cooperative, and limiting the return on capital to make sure that the cooperative primarily benefits member-owners. All these requirements are consistent with the reason that cooperatives were first organized during the nineteenth century: to gain economic power for farmers and workers so that they could form their own capital and own their own means of production.

The second requirement for tax exempt status, "to adhere to the activities for which it was created," means that the cooperative must deliver electrical services though other types of public utility services, including water and broadband internet, have also been allowed to qualify under this requirement as "like organizations." That means rural electric cooperatives can provide many kinds of electrical services, including energy efficiency and demand management, but they cannot sell electrical appliances without being subject to unrelated business income tax, because selling equipment is considered to be outside the activities for which rural electric cooperatives were created.[1] This requirement sets important boundaries around the types of clean energy services that rural electric cooperatives can provide and how those services can be delivered.

The third requirement specifies that no less than 85 percent of the cooperative's income must be derived from its members. Alternatively, this provision means that a cooperative can't earn more than 15 percent of its income from nonmembers or from an unrelated line of business. This test enables cooperatives to have other sources of income and other customers outside its membership, but within specific limits that ensure that the cooperative's focus remains on benefitting its member-owners.

Rural electric cooperatives can also create subsidiary 501(c)(3) non-profit organizations, for-profit businesses, and other types of cooperative utilities, including broadband internet, to meet the needs of their member-owners. While creating and managing multiple subsidiaries can be operationally complex, cooperatives that have the leadership and management capacity to make it happen can use these tools generatively to provide more support and services to rural communities that have been chronically underserved and struggle with persistent poverty.

Public Power Utilities

Public power utilities are units of local of government. There is tremendous variety in how they're organized, including how they relate to other aspects of local government. Keeping this diversity in mind and recognizing the importance of understanding the specific operations of your local public power utility, it's important to appreciate that these utilities are an integral part of larger local government enterprises that may encompass parks, schools, libraries, public safety, sanitation, community development, and other activities.

The fact that local public power utilities exist as a part of the broader municipal enterprise has important beneficial implications, though it's important that their governing bodies maintain focus on the public purposes for which they were created and don't operate the utility as if it were a for-profit business. For example, many enterprise cities have no property taxes and instead use revenue from the utility to fund local government services. Therefore increasing utility revenues means more money to invest in community development priorities like public health and housing. In addition, having the utility under the same roof as planning, zoning, and building code enforcement enables local governments to connect clean energy and efficiency programs with the local policies that create the rules of the road for how buildings and homes are constructed. As a result, public power utilities have the ability to work with colleagues across other city departments to pair efficiency and other clean energy programs with better building energy codes and local jobs training programs such that they all work together for the benefit of the community.

All public power utility business models have the following characteristics in common, according to the American Public Power Association:

- They are owned by and operated for the benefit of the local community they serve.

- They are locally controlled and are typically considered be locally regulated and can therefore nimbly respond to local needs and priorities without seeking permission from state government.

- Their operations are not for profit. Excess revenue is either reinvested in the community the utility serves or returned to residents in the form of lower utility rates. There are no outside shareholders or other owners who take the profits.

- Like rural cooperatives, they have access to low-cost and tax-exempt financing and may also be able to purchase low-cost hydropower from federal government power authorities.

- They are designed to be community focused because public power utilities are ultimately controlled by the residents they serve through locally elected governing bodies.[2]

In some places, however, state and local laws may set limits on the roles that public power utilities can play. For example, in Georgia, the gratuities clause of the state constitution prohibits local governments from investing in private property. While this provision is important as an anti-corruption measure, because public utilities are part of local government, it also constrains the ability of Georgia's public power utilities to invest directly in energy efficiency improvements for homes and businesses. Understanding the unique environment in which the local public power utility operates enables you to design programs and approaches that meet local needs and abide by local constraints.

Generation and Transmission Utilities: Pooling Purchasing Power

Both rural electric cooperative and small-town public utilities pool their purchasing power through generation and transmission utilities, which function as separate and distinct organizations from the distribution utilities that constitute their membership. Organized as either federated cooperatives or nonprofit joint action agencies, depending on the type of utility that composes their membership, generation and transmission utilities develop, finance, build, and purchase generation capacity and oversee the infrastructure that delivers power to their members, who then distribute electricity to individual residential and commercial customers at the local level. Contractual relationships between local nonprofit utilities and the generation utilities that were formed to pool their purchasing power can become an obstacle to building a clean energy future if these contracts lock local utilities into long-term commitments for dirty power (though there has been recent progress on this front).

In 2016, the Federal Energy Regulatory Commission (FERC) found that the Delta-Montrose Electric Association, a local distribution cooperative, had the right to buy power outside of their relationship with the Tri-State Generation and Transmission Association. FERC further barred the penalty fees that Tri-State had levied against Delta-Montrose for buying from third-party power providers. This was a powerfully important ruling that opened the door for cooperatives and other small utilities to increase local renewable energy generation to respond to local priorities and demand. In fact, Rocky Mountain Institute projected that it could result in as much as 400 gigawatts of additional local renewable energy development potential across America's 2,890 cooperative and public power utilities—which could lead to the installation of about 1.25 billion new solar panels.

Profits and New Pathways for Growth

Historically, rural cooperative and small-town public power utilities sold energy to earn revenue and spent it on the people, infrastructure, facilities, and operations that were necessary to generate and distribute their primary product: electricity. Profits, which rural cooperatives refer to as savings and public power utilities call excess revenue, were either returned to the nonprofit utility's member-owners or invested in other public services. In this paradigm, the only means of increasing revenues and profits available to return to member-owners or to plow into other community priorities was to sell more electricity or to increase profit margins by some combination of decreasing costs or increasing consumer pricing.

Either way, if selling electricity were your primary or only product, growth and profitability would depend on sustained or increasing consumer demand for power. By extension, increasing efficiency or expanding the ability of consumers to generate their own power—for example, through rooftop solar—would threaten the growth and profitability of the utility, potentially to the point of threatening its financial health and therefore the mission of the utility itself. Moreover, if selling more electricity at higher margins depended on old technologies like fossil fuels, the growth and financial stability of nonprofit utilities would come at the cost of our climate and quality of life. Given the missions of nonprofit utilities to benefit the communities they serve, selling more polluting electricity from burning fossil fuels would put cooperative and public power utilities in conflict with their fundamental principles.

Rural cooperatives, in particular, have been caught in this trap. Among all types of utilities, cooperatives produce the highest percentage of their aggregate electricity needs from coal and are heavily invested in aging coal-fired power plants dating back to the 1970s, on which they still carry debt.[3] Specifically, the Center for American Progress, the Center for Rural Affairs, Clean Up the River Environment, and co-op association We Own It have estimated that cooperative utilities still carry between $7 billion and $8.4 billion in federal debt tied to coal-fired power plants.[4] In just one example, nonprofit Mountain Association reports that East Kentucky Power Cooperative, a generation and transmission cooperative that serves sixteen local distribution cooperatives across eighty-seven counites, is paying $600,000 per day in debt service on coal plants.[5] Many advocates have called on the executive branch and Congress to pass legislation that would fund loan forgiveness for rural cooperatives so that they can divest from coal and more quickly transition to clean energy. While difficult to quantify, it's reasonable to presume that the public cost of sticking with coal measured in human health, environmental damage, and the economic consequences of preventing rural communities from moving forward faster into the future would be much greater than the remaining rural cooperative coal debt.

Good Growth

Whether or not the federal government steps up to help rural America move beyond coal, there are now an abundance of opportunities to put clean energy to work to support growth in revenue, profits, and community benefits. New clean, distributed technologies mean that rural and small-town utilities have the ability to not only generate and sell electricity from renewable resources but also create new products and services that help their customers reduce household bills while delivering good local economic growth. Implementing clean energy services through the mission-centric lens of cooperative and small-town public power utilities has the further and very important benefit of prioritizing public good over private gain. Depending on an entirely individualistic "every man for himself" approach risks creating a clean energy dystopia in which the wealthy get all the benefits and everyone else just gets what is left—same old disparities, but with new technology.

Creating and introducing new energy services requires capital. While some rural cooperative and public utilities may have deep-enough pockets to fund these investments themselves, many, if not most, do not. Lack of capital is not an insurmountable obstacle, however, because both types of utilities have abundant access to a wide array of public and nonprofit sources of funding and financing to enable them to turn clean energy futures into an opportunity for growth that increases their ability to invest in the places they serve. If any resource is in short supply, it's the people, capacity, and expertise needed to put funds to work in new ways.

Sources of Capital

Rural electric cooperative and small-town public power utilities have access to many, but not all, of the same sources of capital, so it's important to discern which apply to your specific local situation. For example, federal financing programs don't all have the same definition for what constitutes an "eligible borrower," which impacts which nonprofit, public power, and other organizations are able to participate. Some programs define eligible borrowers as cooperative and public power utilities serving communities with a population of twenty thousand or fewer residents. Other programs may set the population cap at fifty thousand residents. Many programs, however, make a special exception for utilities that have active loans with USDA, which could even include public power utilities that serve larger cities. While a tremendous abundance of capital resources are available, there's no "one size fits all" solution. Developing place-based capital and financing strategies is therefore a necessary step toward creating local clean energy futures.

Rural electric cooperatives have two primary sources for accessing the capital they need to finance growth and operations, from building infrastructure and launching new member-focused energy programs to providing zero-interest and very-low-interest loans that sustain rural jobs: the Rural Utility Service within US Department of Agriculture (USDA) and the National Rural Utilities Cooperative Finance Corporation (CFC). The Rural Utility Service was originally created as the Rural Electrification Administration as a part of the New Deal, and from its earliest days provided loans and other support to rural communities so they could build their own electrical infrastructure. The CFC is a nonprofit cooperative financing institution created by its rural electric cooperative members through the NRECA to provide an additional avenue for financing beyond the USDA. Both deliver billions of dollars of public and nonprofit financing supplemented by grants and other support services that are made available through affiliated and aligned partners and programs. Together, they provide a variety of highly tailored and sophisticated financing and funding opportunities to rural electric cooperatives.

Small-town public power utilities may also be able to use USDA financing, and some public power utilities are members of CFC and have access to their programs and services as well. In addition, as part of local government, public power utilities can use tax-exempt municipal bonds to finance investments in electric infrastructure in the same way that cities use municipal bonds to finance other kinds of infrastructure including schools, hospitals, roads and transit, and water and sewer systems. As an indication of the prevalence of municipal bond financing in the United States, there is currently $3.9 trillion in outstanding municipal bonds, according to the Securities Industry and Financial Markets Association.

Other types of innovative funding and financing options may also be available, including financial services from local, state, or regional green banks; philanthropic grants or other types of investments; or impact investments. The availability of these options tends to depend on where you're located, what types of programs you're seeking to support, and whom the programs will benefit. They're not broadly available at sufficient scale—particularly when compared with federal and cooperative financing and the prevalence of municipal bonds—to form the foundation of local clean energy futures, but they can be powerfully helpful to demonstrate new ideas and to take risks that more conventional financing sources might not understand or be comfortable assuming.

Federal Financing: The Rural Utility Service at USDA

History and purpose help describe how and in what circumstances federal tools and programs can be used to meet local needs. The federal financing administration that became the Rural Utility Service (RUS) was created at the same time as rural electric cooperatives specifically to provide long-term loans at very low interest rates to enable the deployment of electrical infrastructure serving rural communities. The scope of financing services offered by RUS to rural utilities subsequently evolved to include water, sewer, high-speed internet, and clean energy infrastructure.

The founding architect of the federal financing enterprise for rural electrification was Morris Llewellyn Cooke, an advocate of scientific management as a means for increasing productivity to improve the prosperity of workers and thereby benefit society as a whole. Cooke had led Giant Power, Pennsylvania's rural electrification program, and used utility sector, demographic, and engineering data to counter the assertion of large private utilities that electricity was a privilege and that it was not economically feasible to electrify rural America. In 1934, Cooke authored an eleven-page paper that quickly became the blueprint of the Rural Electrification Administration (REA), which was created by executive order of President Franklin Delano Roosevelt approximately one year later. Cooke was swiftly appointed REA's first administrator.[6]

Subsequent presidential administrations have not always been so supportive. In 1972, President Nixon temporarily eliminated REA financing by executive order. Thousands of rural electric cooperative leaders organized to fight for the agency, and Congress responded in 1973 with legislation that reestablished REA. Not ten years later, in 1981, David Stockton, who was President Reagan's director of the Office of Management and Budget, proposed sweeping cuts to REA as a part of Reagan's plan to reduce the federal budget. In 1991, Congress abolished the Rural Electricity and Telephone Revolving Fund, though it continued to appropriate REA loan funds annually, with each year's appropriation "scored" based on a calculation of the cost to taxpayers of providing low-interest loans to rural cooperatives. Then in 1992, President Bill

Clinton proposed an end to federal subsidies on REA-insured loans, which would increase interest rates and therefore the cost of capital for rural cooperatives.

Supported by the advocacy of electric cooperatives and their members, Congress has consistently acted to sustain rural access to federal financing. In 1994, through the Federal Crop Insurance Reform and Department of Agriculture Reorganization Act, the Rural Electrification Administration became the Rural Utility Service within USDA and gained the authority to finance water, sewer, and ultimately high-speed internet infrastructure alongside electricity. In subsequent years, Congress authorized additional RUS programs tailored to enable rural borrowers to deploy energy efficiency programs for homes and businesses, install solar on farms, put rural land to work producing clean energy, electrify transportation, and increase resilience with energy storage. Four such programs are particularly important to get to know.

Rural Energy Savings Program

The Rural Energy Savings Program (RESP) was signed into law as a part of the Agriculture Act of 2014 and implemented in 2016 as an on-ramp for rural utilities and other qualified borrowers to launch and expand on-bill programs to help rural households and businesses reduce their energy bills through efficiency. Through RESP, qualified borrowers—which include rural electric cooperatives, small-town public utilities, and nonprofits and other program providers working with qualified utilities—can receive zero-interest loans with a repayment term of twenty years that can in turn be used to install energy conservation measures such as insulation and weatherization, new high-efficiency heating and cooling equipment, and solar and energy storage in homes and businesses served by the utility. RESP loan funds can also be used to replace manufactured homes. Commonly known as mobile homes, older manufactured homes can be outrageously inefficient and costly to heat and cool, making their replacement a significant opportunity for energy efficiency and a better quality of life for the people who live in them. For energy efficiency programs funded through RESP, the cost of energy improvements is paid back on the utility bill of participating homes or businesses over a term of no more than ten years, paid for through the resulting energy savings.

Under the leadership of Luis Bernal, former deputy assistant administrator of the Office of Customer Service and Technical Assistance within the RUS Electric Program and former executive director of the Puerto Rico Energy Affairs Administration, RESP was continually refined and improved from its introduction. As a result, RESP is thriving. The program has grown to more than $50 million per year and is bringing the benefits of energy efficiency and savings to a growing list of communities across the US, through rural utilities in states such as Arkansas, Colorado, North Carolina, Ohio, South Carolina, Tennessee, and Washington.

Energy Efficiency Conservation Loan Program

The 2008 Farm Bill expanded the ability of the RUS Electric Program to make loans to support energy efficiency. As a result, the Energy Efficiency Conservation Loan Program (EECLP) was finalized in late 2013. The types of activities that can be supported by EECLP are similar to RESP, but with several important distinctions related to scale, scope, terms, and what types of organizations are considered qualified borrowers.

While RESP supports more than $100 million in authorized borrowing, EECLP can support billions. Only utilities in rural areas can be approved as borrowers, and EECLP can be used to invest in both energy efficiency programs as well as in improvements to the utility's systems.

The term of EECLP loans is capped at fifteen years or the useful life of the equipment that loan funds are used to purchase, and the borrower pays interest. The comparative scale and different terms of EECLP financing make this program an optimal tool for significantly scaling up existing efficiency programs as well as making larger-scale investments in more efficient equipment and infrastructure, such as installing renewable energy or replacing an entire community's streetlights with high-efficiency LEDs.

Rural Energy for America Program

USDA also provides direct funding and financing for projects to support rural small businesses, farms, and other types of agricultural operations. The Food, Conservation, and Energy Act of 2008 created the Rural Energy for America Program (REAP) from a predecessor program that provided similar support for energy efficiency and renewable energy systems. REAP encompasses both grants and loan guarantees that can be combined to cover up to 75 percent of eligible projects costs. The types of projects that can be supported are extensive and include everything from energy audits and insultation to anaerobic digestors that turn poop from farm animals into biogas for fuel. REAP grants and loans have funded projects in every state and in US territories.

Rural Economic Development Loan and Grant Program

If there were ever any doubt that rural utilities were in the business of community economic development, the creation of the Rural Economic Development Loan and Grant Program (REDLG) in 1987 dispelled it. Fully implemented in 1989, REDLG provides zero-interest loans and grants to promote local economic development and to support and preserve rural jobs. Up to $300,000 in grant funding and up to

$1.5 million in zero-interest loan funding can be used by rural utilities to pass zero-interest loans directly through to qualifying local businesses or to establish a revolving loan fund to provide zero-interest financing to many local businesses. The ultimate application of loan funds can include creating business incubators, providing assistance to nonprofits, investing in health care facilities and equipment, supporting start-up costs for new businesses, expanding an existing business, or paying for technical assistance. Once the initial round of loans has been repaid to the utility, funds can be re-lent in a revolving structure to provide continuing support for community development.

The REDLG program can be combined with other sources of USDA RUS funding to help start up and support the businesses that help build clean energy futures. For example, a rural utility that uses RESP to implement a new energy efficiency program for its members could also disburse REDLG funds in the form of zero-interest loans to local heating and cooling contractors to train their employees and purchase the energy auditing and other equipment necessary to deliver energy efficiency services. REDLG-funded zero-interest loans could further be used to pay for up to 80 percent of the cost of new facilities for a local high school or community college to add training programs for good jobs in the clean energy field.

The National Rural Utilities Cooperative Finance Corporation

USDA programs aren't the only funding and financing programs available to rural and small utilities, or even the most often utilized. Launched in 1969 through the NRECA, the National Rural Utilities Cooperative Finance Corporation was created by 512 founding members as a nonprofit cooperative financial institution to supplement RUS loan programs. As a cooperative, the CFC is owned by its members and exclusively serves their needs. At the time of this writing, it has more than $26 billion in loans outstanding according to Moody's, serves 974 members, and provides financial services to an additional 46 associates. Since its founding, the CFC has both collaborated with USDA, leveraging RUS financing resources, and stepped into the breach when the political winds in Washington shifted and threatened rural cooperative support. For example, when President Nixon temporarily eliminated REA financing, CFC stepped up and provided the loans that cooperatives needed.

CFC membership benefits extend well beyond lending. They include investment services to help manage local cooperative finances, financial forecasting tools and assistance, training and educational events, and other programs that are specifically designed to promote the cooperative business model. The CFC has also participated in the creation of a broader ecosystem of additional cooperative organizations that each play a specialized role in supporting rural cooperatives and their unique business model. For example, the National Cooperative Services Corporation is a nonprofit cooperative that is affiliated with and operated under a management agreement with CFC and offers financing to help rural cooperatives acquire adjacent utility service territories from other utilities.

When you consider that many rural cooperatives serve fewer than a thousand meters and that the average size of rural electric cooperatives is just over twenty-two thousand members, the extensive and interconnected network of financial services and related support is important to provide access and lend additional help, particularly for small cooperatives. CFC financial services are also often regarded as being easier to work with than RUS programs because of the paperwork involved in applying for federal funds and financing. Finally, recall that rural cooperatives serve more than 90 percent of the persistent poverty counties in the United States. They may themselves struggle to build and sustain the organizational capacity necessary to put available resources to work for their communities. In such an operating environment, pooling resources to create shared capabilities is very much in alignment with "the cooperative way."

Tax-Exempt Municipal Bonds

State and local governments use tax-exempt municipal bonds to finance a wide variety of public infrastructure. In just the last decade, these bonds have financed more than $2.3 trillion in new infrastructure investments, including investments in public power inclusive of new sources of electricity generation, distribution, and efficiency.[7] When you purchase municipal bonds, you're essentially lending money to the state, local, or other government entity that issued them in return for a regular interest payment. Interest income from municipal bonds has always been exempt from federal income tax and may also be exempt from state or local income tax if the bond owner lives in the state that issued the bond. Among the reasons that interest income from municipal bonds is exempt from taxes is that it benefits the general public by reducing the interest payments that states and local communities would otherwise have to pay to banks to finance public infrastructure. In short, exempting interest income from municipal bonds means that building schools, hospitals, roads, water and sewage systems, and other utility systems costs less.

The first "green bond" transaction in America used tax-exempt municipal bonds to finance the creation of the Delaware Sustainable Energy Utility. Led by Trenton Allen, then of Citi, who has since founded and leads Sustainable Capital Advisors as managing director and CEO, this pioneering $70.2 million bond offering was announced at the Obama White House's first Better Buildings Challenge summit in December 2011. Projected at the time to support more than a thousand jobs, the Delaware Sustainable Energy Utility has expanded to provide a wide variety of renewable energy and energy efficiency services to businesses and residents.

In principle, investing in municipal bonds allows you to invest in your own community. But investments in municipal bonds are becoming highly concentrated among the wealthiest American households, following a similar pattern in the concentration of asset ownership and wealth that has been observed across many financial markets.[8] The municipal bond market also tends to be the domain of big banks and other large and highly sophisticated financial institutions, though recent tech start-ups have sought to make investing directly in local projects more accessible by crowdsourcing municipal bond investments. San Francisco–based start-up Neighborly was founded by Jase Wilson, who grew up in Maryville, Missouri, a small town served by a rural cooperative utility where his stepfather worked as a lineman—an important touchstone for the company. Before Neighborly closed its doors in October 2019, the company successfully financed a number of local infrastructure projects including parks, libraries, and bike paths. Armed with an understanding of what the company did right and did wrong, or did right at the wrong time, perhaps the next Neighborly will succeed in realigning local municipal bond–financed infrastructure with local investors.

Philanthropy, Green Banks, and Impact Investors

Finally, there are a variety of innovative funding and financing options available through philanthropy, local and regional green banks, and impact investors. While these options don't begin to touch the scale of federal, CFC, and municipal bond financing options, they can be catalysts for building local capacity and expertise and for enabling new ideas that other, more conventional sources of funding and financing may not understand and may perceive as too risky.

Philanthropies can provide both grants and debt capital. When and where it's available, grant funding can be ideally suited to being the "first money in" to support a new idea and help pay for people, training, and other start-up costs for new programs. This kind of support can be critically important to building the local leadership and capacity needed to create clean energy futures. Grants can also be used to provide financial services such as loan loss reserves so that small utilities can move forward with implementing new programs with the confidence that there is grant funding available to fill the gaps in case of unanticipated losses. Depending on individual foundation policies, grants may be payable directly to nonprofit cooperative or small public power utilities, or issued in partnership with values-aligned local nonprofits. In addition, philanthropic debt capital, typically in the form of program-related investments or PRIs, can provide financing with terms that may include lower interest rates or a longer payback term than those available from other lenders.

Similarly, green banks and impact investors may be willing to provide debt financing at favorable terms to support new and innovative programs and ideas. These sources of capital are focused first on their missions in alignment with building clean energy futures, and may have a greater understanding of and appetite for taking risks to support innovation. It's hard to beat USDA RUS loans with zero-percent interest, but there are other benefits of working with financial institutions such as green banks and impact investors that also have expertise, strategic advice, and extensive networks to share.

Worth More than Gold

Combining grants, zero-interest and low-cost loans, and other specialized sources of investment such as green banks, local nonprofit utilities have plentiful access to an extensive funding network that enables them to be early innovators in the deployment of new clean energy technologies as well as to take new technologies to scale.

These funding networks have stood the test of time, evolving and diversifying in response to market and policy dynamics that have unfolded over the past century in order to ensure that local utility needs would be met. The same financial infrastructure equips rural cooperative and small-town public power utilities to realize the opportunities of our current clean energy transformation, and the fact that their business models are owned and democratically governed by the communities they serve empowers them to use clean energy to create local wealth. In the next chapter, we'll begin a detailed discussion of how you can join with your local nonprofit utility to put these tools to work—starting with energy efficiency and incorporating solar power, energy storage and microgrids, electric vehicles, and the broadband networks that are needed to create smart, clean, connected, local, resilient energy systems.

The combined potential of cooperative and public power utilities and all the financial resources the USDA, CFC, and the municipal bond market can muster doesn't matter if there aren't enough folks at the local level to pull those resources down and put them to work. People like you are the most important asset for building clean energy futures, and the history of rural power is filled with the stories of individuals who made a difference.

CHAPTER 5

Energy Efficiency

The First Fuel as a First Step

Often referred to as "the first fuel" since every kilowatt you save is energy you don't need to generate, energy efficiency is also an ideal first step to start building a clean energy future. Every household can benefit from energy efficiency, every community can develop an energy efficiency program, and there are many models to follow. Moreover, the financing tools you'll use and the local technical capacity you'll build will be applicable to other types of clean energy programs, including renewable energy and resilience.

The kinds of improvements that increase energy efficiency (attic insulation, weatherization to seal drafts, new heating and cooling equipment) are not very visible. The combined results, however, are immediate and measurable. From sustaining local jobs to delivering utility bill savings while improving the health and comfort of people's homes, energy efficiency can give everyone access to the benefits of a clean energy future. This is especially important for low- and moderate-income families, whose energy burdens are nearly twice as high as those of their more affluent neighbors,[1] and for communities that continue to suffer housing, health, and wealth disparities due to racist policies such as redlining and segregated housing.

In this chapter, we'll map energy burdens in America and examine the connections between energy, housing, and poverty that have left many families with utility bills that exceed the cost of rent. Once we know where to focus, we'll explore proven approaches for expanding access to energy efficiency that pay for up-front costs with long-term savings. To implement what works, we'll examine multiple examples and learn the steps to energy efficiency, including analyzing the data, engaging the community, training the workforce, and finding the best financing tools. Finally, we'll identify other housing issues you may uncover and how to approach building the additional partnerships you may need to serve your neighbors well.

Hometown Heroes

One woman's journey is central to our story. A small businesswoman and lifelong entrepreneur, Tammy Agard was living in Montana, running a coffee roaster and flipping houses when she heard a Red Cross appeal for volunteers on the radio. It was September 2005, the Gulf Coast was reeling from the aftermath of Hurricanes Katrina and Rita, and Agard knew she had to answer the call. As she shares, "It changed my life."

Hurricanes Katrina and Rita had damaged upwards of one million homes and displaced more than one million Gulf Coast residents, more than six hundred thousand of whom were still displaced a month later.[2] Many homes were destroyed and even more were left uninhabitable by floodwaters that contained sewage and dangerous chemicals and left behind ideal conditions for black mold when they receded.

Agard didn't waste any time. She packed up her things, put her businesses on hold, and headed south. The Red Cross trained her and put her to work delivering meals, but it wasn't long before her experience as a residential contractor put her on the front lines of rebuilding communities. It was now 2006, and the Bush–Clinton Katrina Relief Fund had expanded its support of disaster recovery and reconstruction. Through a program enabled by the Bush–Clinton fund, Agard worked with people who had been victimized by contractor fraud, provided case management, and got them the help they needed to stay in their homes.

The physical damage didn't surprise her, but the electricity bills did. It was not at all unusual for people to be paying $400–$500 per month to cool their homes and keep the lights on. This was not a product of Katrina, but rather the result of housing with no insulation, old and inefficient heating and cooling equipment, and air leaks equivalent to a massive hole in the side of a home. For these people, many of them seniors on fixed incomes, paying the bill to keep the lights on came at the cost of other essentials, like groceries and medicine. Subsequent research studies would show that their stories were not the exception but the rule.

It was during this time that Agard met Johnnie LaCaze, a contractor and lifelong Mississippian, who would eventually become her business partner. Their work together on the Gulf Coast would take them next to Little Rock when the Clinton Foundation brought the energy efficiency retrofit programs that had helped low- and moderate-income households in and around New Orleans back home to Arkansas.

Led by Martha Jane Murray and the Clinton Climate Initiative, the HEAL (Home Energy Affordability Loan) program had been directly inspired by a program Murray and her husband launched for employees at their shoe factory in Wynne, Arkansas. An architect and long-time leader in the green building movement, Murray made completing an energy audit and retrofit on the factory a first order of business. She wanted to make the same kinds of energy savings available to the company's employees, so created HEAL with a forty-thousand-dollar

Tammy Agard and Johnnie LaCaze, cofounders of EEtility, joyfully building out their first office, 2014.

revolving fund to offer zero-interest loans to employees for retrofitting their homes. Early program results showed energy savings averaging 30 percent, which put money back in people's pockets by reducing their electricity bills.[3] The program Murray pioneered to serve her colleagues was adopted by the Clinton Climate Initiative, and Agard relocated to Little Rock in 2009 to lead it.

While HEAL helped nearly 750 households cut their energy bills by a total of $625,000 per year,[4] with Hillary Clinton's presidential campaign on the horizon, it wasn't clear if or how the program would continue to grow. So five years later, in 2014, Agard and LaCaze founded EEtility and picked up where the HEAL program left off.

Depending on grants and loans as HEAL had wasn't scalable, so Agard and LaCaze had to find a better way to expand access to energy efficiency. Through Dr. Holmes Hummel of Clean Energy Works, a thought leader on energy finance, they learned about Pay As You Save (PAYS), an on-bill approach to financing energy efficiency developed in Vermont. It was just what they needed. While HEAL was delivered through employers to their employees using loans, PAYS could be delivered to everyone through local utilities using utility bills.

Agard saw the potential of the PAYS model and brought it to Ouachita Electric Cooperative, a small rural utility in south central Arkansas. The result became HELP (Home Energy Lending Program) PAYS Arkansas, which in its first year helped more than two hundred households save more than sixty dollars per month on their bill on average.[5] As Mark Cayce, Ouachita's CEO observed, "It's changing lives, because if you need new heating and air equipment, it's a major purchase. It can be as expensive as buying a car. If you already have to buy a car, or do something else to your house, there's just not enough cash to make ends meet. This program gives people the opportunity to put more money toward other things in their lives. Maybe it's medicine, maybe it's sending their kids to school, or something else. But it eases their burden a little bit in life."[6]

The eight years from post-Katrina New Orleans to Ouachita were filled with important lessons that not only shaped Agard and EEtility but also informed how communities across America are approaching energy efficiency. Moving from limited grant-funded approaches, through early experiments with loan-based initiatives like HEAL, to on-bill energy efficiency programs that are paid for with savings without consumer debt represents an important evolution for the energy sector as a whole. Agard's presence as a purpose-driven entrepreneur every step along the way was pivotal because she was able to take what she had learned and put it into practice through EEtility. Without her steady implementation, we might not have a program model that any local utility, anywhere, can implement today.

Relieve Energy Burdens

As in the Gulf Coast and rural Arkansas, everywhere across America, lower-income households carry disproportionately high energy burdens. Understanding the extent of these disparities and why they exist is essential to creating energy efficiency programs that improve the financial well-being of families and the economic vitality of the community as a whole. If you do not design your approach to meet the needs of people with high utility bills living in old houses without enough income to make ends meet, you will end up with an energy efficiency program that helps only the affluent, and the results will worsen economic disparities instead of alleviating them.

Low income is typically defined as earning 50 percent or less than the area median income (AMI) for the community where you live, and moderate income is typically defined as earning less than 80 percent but more than 50 percent of AMI. In my home state of Georgia, for example, a family of four bringing home $55,500 per year or less in combined household income would qualify for Georgia's Weatherization Assistance Program for low- or moderate-income households. (If you think of yourself and your family as middle income, you may be surprised at what middle income means in your state and community. The Pew Research Center has a straightforward online income calculator.)

Energy burden is defined as the percentage of total household income spent on utility energy bills, and low-income households pay higher bills because their homes and appliances tend to be older and less energy efficient. Moreover, households living on low and moderate incomes lack the discretionary funds to buy new appliances or pay for energy improvements like insulation out of pocket. The national statistics are alarming, but it may be even worse where you live. The average US energy burden for low- and moderate-income households is 14.5 percent, but the average energy burden exceeds 20 percent in seven states (see table 5.1).[7]

Table 5.1. The 20 Most Energy-Burdened States

State Abbreviation	Low-Income Energy Burden
AK	42.4%
ME	40.4%
VT	27.2%
MS	26.7%
HI	23.1%
SC	22.0%
AL	20.9%
NC	19.8%
NH	19.7%
GA	19.4%
RI	19.4%
KY	19.1%
AR	18.9%
PA	18.7%
CT	18.5%
WV	17.8%
NY	16.3%
AZ	16.2%
MA	15.6%
NM	15.2%

Table 5.1. These twenty states have the highest low-income energy burdens in the US. (Derived through aggregation of county energy burdens for each state. Source: Moleka, 2021.)

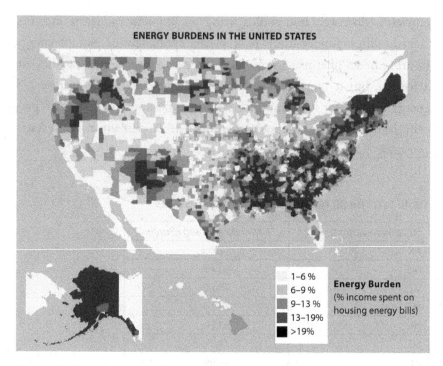

Geographic distribution of energy burdens (percentage of household income spent on household energy bills). (Source: National Renewable Energy Laboratory; image from NREL data visualization.)

Energy burdens hit rural communities even harder, especially in the South and the upper Northeast. According to a 2018 study from the American Council for an Energy-Efficient Economy, rural low-income households experience higher energy burdens than low-income households in metropolitan areas. In some regions, more than a quarter of rural low-income households have energy burdens of 15 percent or more, and across the board, the elderly, communities of color, renters, and families living in manufactured homes suffer the worst.[8] For these households, energy efficiency is not a luxury but a necessity, though it is often out of reach.

Improving Efficiency, Paid for with Savings

The combination of rural poverty and high energy burdens makes energy efficiency more than a way to build clean energy futures. It's an important and very relevant tool for alleviating poverty. Sustaining the level of investment that is necessary over time, however, requires a fiscally sustainable approach, just as Tammy Agard found through her work in Arkansas. As one measure of the persistent gap between the need for energy assistance and the funding that's available to help pay people's bills, we can look at the federally funded Low Income Home Energy Assistance Program (LIHEAP). Approximately thirty-five million US households are eligible for LIHEAP, nine out of ten of which are home to a veteran, young child, an elder, or someone with a disability. Although annual funding for LIHEAP exceeds $5 billion, there's never enough to meet the need. In 2019, only one in five households who qualified for help received any assistance.[9] Moreover, because LIHEAP pays the bill but doesn't address the root causes of high energy bills, it will never be sufficient on its own.

In places served by investor-owned utilities (IOUs), state regulators may mandate that the monopoly utility derive a certain percentage of its new electricity generation requirements from energy efficiency or may mandate that a certain amount of money be invested in efficiency improvements for utility customers in specific rate classes, which may or may not be linked to household income characteristics. Whether such mandates view energy efficiency as a fuel source and set a percentage target or simply set a dollar-value goal, the impact is the same—the utility must direct some portion of its revenue toward paying all or a portion of the cost of energy efficiency improvements to people's homes. While many energy advocacy groups are accustomed to working for policies that force IOUs to allocate a few million of their billions of dollars in revenue to residential energy efficiency, it's different for small nonprofit utilities that are owned and controlled by the people they serve. As we have learned, rural cooperatives and public power utilities exist to deliver power and community benefits, so energy efficiency programs that relieve energy burdens can be seamlessly incorporated into their core business.

Unlike loan programs that depend on consumer debt or residential Property Accessed Clean Energy programs that depend on property tax assessments for repayment, on-bill financing programs like the PAYS program structure that Agard and her company EEtility use are paid for through energy savings on the utility bill. As a result, anyone who pays a utility bill, whether they rent or own their home, is able to participate. Moreover, the repayment obligation follows the meter, not the resident, so the potential that a participating resident might move before the end of the repayment term doesn't present an obstacle.

Because reducing utility bills is the core purpose of this model, its implementation is based on a rigorous home energy assessment process that identifies the list of energy conservation measures that will save more money than they cost over time, customized for each individual home. Payback models are also specific for each utility service territory, so the approach accounts for variations in energy pricing and rate structures from place to place. The program is designed so that participating residents experience immediate savings while also paying back the cost of efficiency improvements on their utility bill by sharing the savings between the resident and the utility. The term of repayment is typically ten to twelve years, depending on the local market conditions, and the resulting revolving structure of the program's financing means it's fiscally sustainable for the long term.

This approach aligns with nonprofit utility business models, and enables local communities to prioritize grant-funded energy efficiency improvements for residents with the greatest needs. As you'll recall, local nonprofit utilities exist for the public good, so sustaining utility revenues

HOW PAY AS YOU SAVE® WORKS

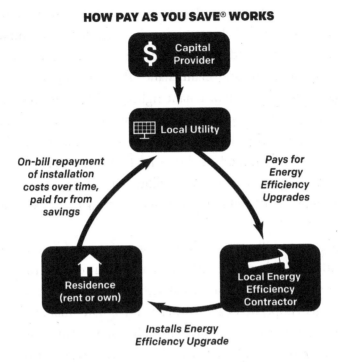

How the Pay As You Save on-bill financing program for energy efficiency works. (Original graphic.)

helps sustain the community. While energy efficiency may reduce the volume of electricity a small utility can sell, it creates new sources of revenues from improving energy efficiency as an energy service, which offers additional mission-aligned benefits for the community. The risk of long-term utility revenue loss is low when weighed against the potential for new energy services delivered through similar "paid for with savings" models, which could include energy resilience or electrifying vehicle fleets like school buses—both of which we'll address in later chapters.

On-bill programs should also be combined with other types of energy efficiency initiatives that are supported by federal, state, philanthropic, or other grant sources because they will not be sufficient to relieve all energy burdens on their own. The energy assessments that are part of any PAYS program will likely reveal deep housing disparities and unsafe and unjust housing conditions, which may require other strategies, including major repairs and maintenance for owner-occupied homes, strengthening housing policy, tenant rights, and landlord responsibilities. While grant funding goes up and down over time and can't be relied on for long-term sustainment, grants are optimal for addressing housing disparities that markets alone can't heal, and for supplementing smaller-scale program start-up costs, filling gaps, supporting research, and addressing related home repair needs like leaky roofs and broken windows.

Offering an on-bill energy efficiency program may be new to many local utilities, but it's a well-established approach to delivering energy services. In fact, the Environmental and Energy Study Institute has identified fifty cooperatives across twenty-three states that offer or have recently offered similar on-bill programs, though their innovations haven't yet been widely recognized or celebrated. This wealth of rural leadership creates the opportunity to replicate an energy efficiency solution that's already working.

Reduce Bills, Save Money, Improve Lives, Support Jobs

For cooperative member-owners and residents, reducing electricity bills means affordability and enhanced quality of life, including more money to spend on household necessities and a more comfortable home. Comfort has value. If you are accustomed to not using central heat or air because you know you can't afford the energy bill, you know how miserable it can be to try to work, rest, study, or sleep in the blazing heat or freezing cold. Living without air-conditioning in the summer or heat in the winter can also be a serious health risk to individuals as well as communities. Excess heat and humidity can lead to heart failure for people with heart disease and aggravate other serious chronic conditions, such as asthma. These same diseases and other underlying conditions, including diabetes, are more prevalent among people living with lower incomes, who are the same people living with the highest energy burdens in housing that is also more likely to be located in a neighborhood with disproportionately high pollution that makes people sick. It's a vicious web of environmental factors that traps people in poverty and chronic disease.

A growing body of research is examining these and other social determinants of health, and the emerging bottom line is that the best predictor of life span in the United States is where you live. Visit the Robert Wood Johnson Foundation's website, enter your zip code, and see for yourself.[10] While improving your home's efficiency isn't a surefire prescription for better health and a longer life, it creates conditions that support health and well-being for the long term.

The dollars invested in energy efficiency also support local jobs and small businesses, especially when local utility programs are designed specifically to do so. In 2020, the International Energy Agency estimated that 14.8 jobs are supported per $1 million invested in energy efficiency; by contrast, every $1 million spent on unabated gas-fired power supports a meager 4.4 jobs.[11] Broad economic job creation estimates can be a little abstract, so let's bring it down to the local level. Installing a new air-conditioning system means hiring a local technician, just like weatherizing a home and installing insulation means hiring a local contractor. Either way, the work is fundamentally local and can't be offshored. Moreover, materials and equipment are purchased from local retailers or wholesalers, often using American-made products. Investing in local business and local jobs through efficiency instead paying for distant dirty energy is a good deal.

The results identified in national studies are visible at the local level too. In a September 2016 report from Ouachita Electric Cooperative, Clean Energy Works, and EEtility, the HELP PAYS program reported participant energy savings that exceeded 30 percent. Including single-family homes, multifamily properties, and two local businesses that enrolled in the program, the program reached 149 member-owners.[12] Similarly, Roanoke Electric Cooperative in North Carolina, which also works with Agard and EEtility, reported that the average cost of energy efficiency improvements in participants' homes was $6,900, yielding an average savings of $120 per month, based on the initial 75 participants in its on-bill Upgrade to $ave Program. Even after the monthly fee or "tariff" was added to participating residents' bills to repay the cost of the installation, residents retained $58 per month on average in savings.[13]

Follow the Leaders

To get going on a new energy efficiency program for your community, follow the leaders. Understanding their approach, why they chose it, and what they learned in the process can help you to align the value of an energy efficiency program with the priorities of your community. That's exactly what I did.

In 2016, Groundswell began working on a solar project to serve LaGrange, but what started as a conversation about renewable energy soon turned toward energy efficiency. No matter whom we talked to, the first thing on people's minds was reducing bills for those who struggled to make ends meet. Patrick Bowie, the local utility director, explained it to me in one of our first meetings. While the City of LaGrange Utility had very low electricity rates, many residents living on fixed or low incomes had high bills because their homes weren't efficient. Homeowners didn't have the financial resources to invest their own money in efficiency improvements like newer and better air-conditioning systems, and renters didn't even have the option but were stuck paying the bill.

As Mayor Jim Thornton noted, high bills for families with low incomes had long been a top concern, but the City of LaGrange was very limited in what it could do to help, in part because the Georgia constitution prohibited local governments, and by extension utilities operated by local governments, from paying for improvements to private property. Apart from the handful of homes that received weatherization from federally funded programs each year, there wasn't a clear solution. Finding one together quickly became our goal.

A year later, I was working to pull together a panel to discuss energy equity for GreenBiz's VERGE conference, and I was hoping to find a rural community leader to join us. My friend and former colleague Mike Kruger suggested I reach out to Mark Cayce from Ouachita. He had just been recognized with an award from the Smart Electric Power Alliance for an energy efficiency program that was helping local low-income residents. I'd never heard of Ouachita or Mark, but I picked up the phone that very afternoon. As soon as we started talking, I realized I'd found not only the perfect addition to our panel but a solution that could work for LaGrange too.

I learned everything I could from Mark and introduced him and Ouachita's HELP PAYS program to LaGrange. Mark very graciously agreed to visit and present his experiences to the city council and members of the community during a trip he had planned to Georgia. His generosity with his time and expertise, and his willingness to share his experience with local leaders, helped give LaGrange the confidence to move forward toward emulating Ouachita's success.

Put People First

Gaining the support of local leaders is just part of the picture. You've got to spend time understanding the needs and priorities of the people you want to serve. It makes the difference between doing something to, for, or with the community. Working with someone implies that they're engaged and on board with what, why, and how something is happening. Even if you want to do something for people that is meant to be helpful—like implement an efficiency program—leaving them out of the design and planning process puts your intentions at risk. The people and communities you hope to serve are the ultimate experts in what they need, and their wisdom and insights are central to good outcomes.

In LaGrange, hearing from local community members was an absolute must-do for the mayor and city leadership. Over numerous breakfast and lunchtime conversations and untold gallons of coffee and iced tea, we learned several very important things. Like many small towns, most of LaGrange's residential rentals are single-family homes. Many of them are located in the community's mill villages, which are vernacular architecture in southern towns that once depended on the textile industry. Organized into neighborhoods around the textile mills and incorporating churches, schools, and neighborhood-scale retail, the mill villages are made of up modest wooden houses built up through about 1940 by the mills for their workers.

Often, these homes lack central air-conditioning and insulation, and too many of them are in poor repair, particularly those used as rental properties. While no one we spoke to was able to characterize just how widespread these conditions might be, everyone had at least one story about someone they knew or a home they had visited with active roof leaks, water leaks in the bathroom, mold, weak or rotted flooring, or worse. Homeowners were often elderly and lacked the resources to catch up on decades of deferred maintenance. Families who rented their home didn't always know who their landlords were or how to reach them beyond the post office box where they sent the rent check.

Everyone we spoke to expressed concern that we would need a plan to keep families from becoming houseless when we inevitably found living conditions that were beyond the ability of an energy efficiency program to address. The community members we met with also shared a universal concern that the same landlords who would allow those kinds of conditions to persist would cash in on energy efficiency improvements like new air-conditioning equipment by going up on the rent.

Two other priorities emerged. While there was a lot of excitement about a program that would help bring down bills, stakeholders wanted to make sure that we focused our outreach on those residents most likely to benefit from the program based on their energy usage. Many of the people we would be working with had suffered far too many disappointments and traumatic experiences already, and no one wanted to make any promises we couldn't keep. In addition, everyone shared a desire to work with local Black-owned businesses to maximize the impact of energy efficiency investments in the broader economic life of the community.

We listened and adapted the program's design to better align its implementation with what we had learned. Engagement with other local programs to provide wraparound services to residents whose homes needed more than just efficiency, publicly signed landlord pledges not to raise rents, and partnerships with Black-owned businesses that could also offer workforce training were three of the localized improvements we incorporated into the plan. When we reported our findings and resulting program updates back to the LaGrange City Council, we got the green light to move on to the next step. Based on the PAYS model, we were ready to launch the LaGrange SOUL (Save On Utilities Long-Term) program.

Energy Usage, Housing, and History

To develop a detailed plan for implementing an on-bill energy efficiency program, it's critical to understand energy usage patterns for residences in your community. Your local utility can do this analysis internally, or it can partner with a values-aligned organization that has energy and data expertise.

Identifying the median energy consumption for residential customers is the first step, which can be determined by analyzing a year of residential energy usage data. Equipped with this information, you can quantify the number of residential customers with the highest energy usage in order to project the potential size, and therefore the potential cost, of your program for planning and budgeting purposes. Depending on what other types of data may be available, you can also analyze energy usage data against tax records, community development data, census tract data, and other sources of information to get a reasonable sense of energy usage per square foot and how many homeowners versus renters might be program participants. If your community prioritizes serving households with low and moderate incomes, census tract data can help you focus your efforts where energy burden relief is needed the most.

Researching the history of your community can help make your energy efficiency investments reparative for people and neighborhoods that have suffered from disinvestment and deeply rooted systemic racism, the impacts of which are widespread and long-lasting. Beginning in the 1930s, a Federal Housing Administration policy known as "redlining" denied mortgage insurance in areas in or near African American neighborhoods. The term itself originated in the New Deal era, when mortgage institutions and the Federal Housing Administration created color-coded maps that indicated areas where African Americans lived in red to let appraisers know that those areas were considered "too risky" to issue mortgages. These racist practices were banned by the Fair Housing Act in 1968, but their impacts are still felt today. Redlining is among the reasons that Black wealth in America is only about 5 percent of white wealth, and that Black homeowners still suffer lower home appraisals than white homeowners, which further erodes wealth. Focusing energy efficiency program and other investments in service to our neighbors who have been hurt by these practices can be reparative.

In LaGrange, our data analysis found that 2,978 residential customers had annual electricity usage that was at least 20 percent higher than the median. Of those, 1,398 or 47 percent lived in census tracts characterized by low household incomes. To make the data easier to understand, the Groundswell team mapped residential energy usage, removing street addresses to maintain the privacy of residents. Visualizing the data with mapping tools showed a clear picture of geographically concentrated energy burdens in LaGrange's former mill villages, particularly in the Calumet Park community, which had once been the segregated mill village for Black and African American residents. Energy burdens and the area's history of racial segregation were fundamentally intertwined.

Workforce Development

In order to invest the full value of your community's energy efficiency program into the community's economic well-being, you'll need to understand if your local contractors have the skills that will be necessary to implement it or if they will need additional support. Providing workforce training, in addition to making sure that jobs stay local, can help people find pathways into careers with good pay in the building trades. There is ample opportunity, well beyond energy efficiency alone. Nationally, according to the US Chamber of Commerce Construction Index, 83 percent of contractors have trouble finding skilled labor. Moreover, the average age of skilled tradespeople is getting older, which means there will be even more demand for younger tradespeople as the previous generation retires. Any way you look at it, the building trades need more trained people to join their field.

Implementing an energy efficiency program requires home energy assessors, heating and cooling technicians, electricians, weatherization contractors, and sometimes plumbers and roofers as well. There are ample resources available through national training programs such as those offered through the Home Depot Foundation and the US Department of Energy. Local programs may also be accessible through area community or technical colleges, or you can partner with local contractors to help tradespeople develop the skills they need as a part of their jobs.

For example, Ed Gresham is a Black business leader whose company, Home Diagnostic Solutions, is the primary weatherization partner for LaGrange SOUL. Ed himself is certified in energy efficiency through the Building Performance Institute as a Building Analyst Professional and through RESNET as a certified Home Energy Rating System (HERS) rater. He has organized and delivered a host of workforce training programs across Georgia as a leader in the community, and he similarly committed his company to training people for opportunities in construction through his work in LaGrange.

Sources of Financing

You've gotten your community's buy-in, quantified the need, your plan is focused on the people, and you've got a go-to contracting team. Now it's time to identify a sustainable way to pay for the program. The availability of financing will not be an obstacle, though accessing it will take commitment, time, and patience with the paperwork. Identifying the right resource will primarily depend on the local population and whether your community is served by a cooperative or a public power utility.

Debt financing, including zero-interest loans, is widely available to start and sustain energy efficiency programs. USDA's Energy Efficiency Conservation Loan Program (EECLP) and Rural Energy Savings Program (RESP) have been widely used to implement on-bill energy efficiency programs—including those offered by the Ouachita and Roanoke electric cooperatives—because both are specifically designed to be used for revolving funds. RESP is a particularly good fit for small utilities serving communities with fewer than fifty thousand residents. Loan terms are twenty years at zero-percent interest, which enables utilities or their partners to use RESP to establish new programs and build momentum without the pressure of paying interest. Established and larger programs can grow into using EECLP, which has a low interest rate but a much higher potential loan size. Similarly, the NRUCFC and local and regional green banks may serve as sources of financing. Green banks, in particular, are expert in energy programs and technologies and can also be helpful strategic partners in program implementation.

Municipal bonds can also be a source of financing. As you may recall, the very first green bond deal in America established the Delaware Sustainable Energy Utility, which got started by offering energy efficiency programs. According to the Climate Bonds Initiative, the US green bond market grew to $51.1 billion in 2020,[14] demonstrating how much more accessible and widely used bond financing now is for clean energy, efficiency, and other sustainability-oriented projects and programs.

LaGrange chose to use available community development funds to support its first year of program implementation, aligning the source of funds with the program's exclusive focus on serving low- and moderate-income residents. Local community approaches have also included pairing program financing with philanthropically supported loan loss reserves. Loan loss reserves can give community leaders additional confidence by making sure that program losses, if there are any, will be paid for without impacting the overall function of the community's utility. This can be particularly helpful to support innovative approaches that challenge utilities to stretch beyond their comfort zones.

Implementation Partnerships

There's no need for you to start from scratch. There are a host of businesses and nonprofit organizations, many of which have experience delivering on-bill energy efficiency programs, that you can work with to bring energy efficiency to your community. If your hometown, state, or region doesn't have a go-to energy efficiency leader, reach out to one of the local utilities that already has experience and ask for a referral. There are lots of people who can help you chart a path, just as Mark Cayce of Ouachita helped me.

Be mindful of the importance of values alignment when choosing partners. There are many organizations with extraordinary expertise in the technical aspects of energy efficiency, but with little or no experience working with residential customers or customers with low incomes who have different needs. If a potential partner's primary focus is corporate or institutional clients, they may not be the right match for your community.

Like Ouachita, Roanoke, and a growing list of utilities across the country, LaGrange works with Tammy Agard and EEtility on technical implementation, and partnered with Groundswell to lead community engagement and local partnerships. The community formally launched LaGrange SOUL in the summer of 2020, three years after we first began searching for solutions, following an extensive period of analysis and engagement. Many roads lead back to Agard and the long-lasting impacts of her life-changing experience helping people in the aftermath of Katrina. I could not be more grateful to work with her and her team to serve my hometown.

More Needs Than Energy Efficiency

Because a fundamental part of any on-bill energy efficiency program is a home energy assessment, implementation will reveal much more about your hometown's housing conditions that what is visible from the street. In low-income communities where residents may not have enough money to pay for roof repairs and other maintenance, you may find that your neighbors need much more help than what an energy efficiency program can deliver. And you will see just how poorly many landlords maintain properties where people live. Be prepared to find additional local partners that can help fill the gaps. Local housing agencies, national organizations with a local presence like Habitat for Humanity, community foundations, and other service organizations may all be good additions to your team.

For example, 95 households signed up for SOUL during the program's first six months of operation. While we knew some homes would have maintenance issues based on what we'd learned by listening to the community, we were not prepared for the extent of what we found. Among the residences that signed up and received an initial energy assessment, 53 percent had maintenance needs that prevented them from moving forward in the program. Of those, 59 percent required major repairs including roof leaks, broken windows, unvented gas appliances, sewage leaks, and rotted stairways. While SOUL focused on energy efficiency, its implementation had revealed the extent of housing disparities in the community, underscoring the need for the broader reparative action on housing long championed by Zsa Zsa Heard, the dynamic CEO of the City of LaGrange Housing Authority and West Georgia Star.

For those residents who were able to move forward with energy efficiency improvements, results were strong. Within the first few months, eight homes had completed the program with at least forty-five days to track energy efficiency results. The average value of these energy efficiency improvements was $4,796 per residence, leading to an average of $555 per year in projected energy cost reductions and about $120 per year in net savings to participants. Residents who previously received among the highest electricity bills experienced the greatest savings. One family of elderly homeowners who received more than $11,000 in energy efficiency improvements for their home, including a new heating and air-conditioning system, saw their monthly bills cut in half. As Chad Cooper of LaGrange Housing Authority observed, these results positioned SOUL as a helpful tool for housing preservation, particularly where SOUL could help homeowners afford repairs that sustained their homes and reduced their energy costs for the long term.

A Necessary Foundation

Energy efficiency can accomplish more than reducing utility bills and energy burdens. It has the potential to relieve poverty, improve living conditions, and enable people to enjoy a healthier and more comfortable home. Implementing an energy efficiency program will reveal housing and related health and wealth disparities, particularly for people who rent their homes. Visible throughout the country and highlighted in national statistics, these disparities result from generations of racist systems, disinvestment, and abuse. It is our generation's privilege to be alive at a time when we can use energy efficiency as a reparative investment to help do justice.

Delivering an energy efficiency program will also build up the technical capabilities your community will need to press into its clean energy future. Developing workforce skills like energy assessment and weatherization, expanding the capacity of your local utility to deliver energy services, and implementing new financing tools all create a strong foundation for what comes next: renewable energy.

In the next chapter, we'll discuss solar power—its alignment with the abundant resources of rural America, how it's developed and financed, and how the value it creates can be deeply connected to the economic well-being of the place you live.

CHAPTER 6

Solar

The Last Crop

Whether it's installed on a rooftop or covering thousands of acres of fields, solar power has become part of America's everyday energy landscape. Driven by corporate renewable energy commitments and clean power goals and enabled by falling installation costs, there is already enough solar energy capacity installed nationwide to power 18.6 million homes—and that's just the beginning. Solar installations in the United States are expected to grow by more than 400 percent over the next decade based on market trends, which translates to more than 1.6 million acres of new solar farms and a tremendous economic opportunity for rural communities.[1] That opportunity is only expanding as new and bolder policies emerge, such as the recent US national goal of achieving a 100 percent carbon-free US power sector by 2035.

While national policy goals don't dictate how local communities build clean energy futures, they do set a direction and pace. To date, public power utilities get 20.6 percent of their power from renewable sources, investor-owned utilities get 11.2 percent, and cooperative utilities get just 2.4 percent.[2] To meet the mark, we will need to build a whole lot more solar, wind, and other carbon-free energy capacity at a much faster clip than the growth that's already projected.

While rooftop solar installations on homes and businesses may be most familiar to us, large-scale installations account for the vast majority of solar's growth. Typically, solar developers lease land for at least twenty years, install solar panels, and sell the resulting electricity to the local utility or to local businesses and residents, depending on what state laws and regulations allow. The value of a twenty-year solar lease is why so many farmers call solar "the last crop." A 2017 report from the North Carolina Sustainable Energy Association found that the typical annual solar lease payment for farmland ranged from $500 to $1,400 per acre—about seven times that of an agricultural lease. A 100 MW solar project on five hundred acres would yield at least $250,000 per year in lease income for a generation—plus the family would still own the land at the end of the lease.[3] Land leases that build wealth and help preserve multigenerational land ownership are just one dimension of the economic benefits of solar power to local communities. Additional sources of income and other value streams include dual land use that combines energy and agriculture, the potential for lower energy costs, increased local property tax revenue, increased competitiveness, and new energy industry and construction jobs as well as electricity revenue, tax credit revenue, and other types of incentive revenue if the solar project is locally owned.

You can help connect the rapid rise of solar with the economic revitalization of your hometown. What's possible and how to realize it depends on the energy laws and regulations of your state, which define your solar market and in turn drive how to use your local sun and land resources. The type of utility that serves your community shapes how you connect the resulting value to your local economic development priorities, then there are three scales of solar—large utility scale, community, and rooftop—that you can put to work to meet your needs.

The tremendous variety in how these factors come together does mean there's no single recipe for developing a solar strategy for your community. Adding to the complexity, energy policy, which is a fundamental building block of energy markets, is rapidly evolving and expanding the field of opportunity. While these dynamics can be as frustrating as they are exhilarating, taking an entrepreneurial approach to understanding your state and local solar market, being prepared to change policy if and when it's necessary, and getting familiar with examples from similar communities will illuminate your vision and help you to create a plan. Importantly, you also have thousands of innovative colleagues working toward similar goals in the places they live, myself included, who can be helpful sources of inspiration, information, connection, and support.

Connecting State Policy and Solar Markets

Every successful solar strategy begins with an understanding of whether your state allows anyone other than the utility to sell electricity and whether your state offers solar or other renewable energy incentives. The answers to these two questions will define how you connect the value of solar to local community development priorities and will shape how big or small your solar projects can be, who can own them, and if and how much savings are available to local businesses and residents. Two examples—one from a regulated state with no incentives and one from a deregulated state with specific renewable energy goals and solar policies—illustrate why and how.

Regulated markets with electricity monopolies and no renewable energy incentives can support healthy solar markets, but they look and operate differently in states that have strong incentives and policies that encourage competition. Georgia Public Service Commissioner Tim Echols calls it "Red State Solar." Market-driven by large customers and built at or below avoided cost, the approach has enabled southeastern states like Georgia to push up into the Solar Energy Industry Association's (SEIA's) Top Ten solar states list.[4] Its success has been dependent in large part on surging renewable energy demand from large corporate and institutional buyers whose long-term electricity contracts, which are commonly known as power purchase agreements, stand behind thousands of acres of solar panels. The US military was among the earliest leaders whose renewable energy purchases helped to make the market. In January 2012, President Obama announced the navy's commitment to deploy 1 gigawatt of new renewable energy to power its US bases in his State of the Union address. Just three months later, in April, the army and air force joined in for a combined goal of deploying 3 gigawatts of renewable energy by 2025 to support the mission and make US military operations more resilient.[5] These bold commitments helped get solar off to a running start in states with multiple military bases. For example, Georgia Power's first large-scale renewable energy projects included three 30 MW solar arrays serving military bases, the first of which was announced in August 2014 and included arrays at Fort Benning, Fort Gordon, Fort Stewart, and Kings Bay.[6]

Red State Solar economics doesn't assign any value to the climate or public health benefits of solar versus the costs of polluting sources of energy that are borne by the public in addition to whatever direct costs show up on the utility bill. Under this approach, big projects deliver economic savings while smaller projects struggle to compete. Big solar takes a lot of land—from four to five acres per megawatt as a practical matter, though technological advances in solar efficiency mean less land will be required to produce the same amount of energy in the future. As a result, the availability of large tracts in rural communities is a competitive advantage for attracting solar infrastructure investment.

A second important implication of Red State Solar is related to competitiveness and market access. Because states like Georgia have regulated markets, solar project owners have to sell their power to the utility and are forbidden by law from selling retail electricity directly to residents and businesses. Having just one or only a small set of utility customers means utilities are able to set the price of solar electricity instead of the price being determined by an open and competitive retail electricity market. These regulatory and market constraints also tend to favor large solar projects with economies of scale because they can meet utility electricity cost requirements.

In states that set specific renewable energy goals and allow solar project owners to have direct access to sell electricity to residential and business customers, the economics, and the opportunities, are broader and accommodate smaller local solar projects. Thirty states, the District of Columbia, and three US territories have adopted renewable portfolio standards (RPS) that set specific goals and schedules for what percentage of the state, district, or territory's electricity must come from renewable energy. Utilities that are subject to state RPS goals are required to purchase Renewable Energy Certificates to verify that they are meeting state targets. RPS goals thereby enable the state's energy market to value clean energy more highly than polluting sources of energy, which makes sense because clean energy has more benefits and fewer costs. Some states also set specific goals for solar deployment within their RPS and may create additional incentives for expanding solar access to households with low or moderate incomes. Often referred to as "Solar for All" policies, these incentives are particularly important for helping to restore communities that have suffered from disinvestment and redlining, which among other consequences often leaves them with older and less efficient utility infrastructure that costs more to upgrade in order to install new energy technologies, including solar.

The Rockford Solar project in Illinois is a great example of how state RPS and Solar for All policies can be put to good use to deliver local value. Located on a filled rock quarry that used to send local residents running to shut their windows when dynamite shook the neighborhood and sent dust flying, the site now hosts 6,600 solar panels that are delivering electricity and savings to more than five hundred local residents with low and moderate household incomes through the Illinois Solar for All program. Working with a values-aligned solar developer enabled Rockford to host a solar project that used state incentives and available land to meet the needs of the local community. Trajectory Energy founder and managing partner Jon Carson grew up on a fourth-generation dairy farm in western Wisconsin. A passionate advocate for rural communities, Jon worked closely with local leaders through the Rockford Solar for All Coalition to design the project together with the people it would serve. The value of the available Illinois Solar for All incentives helped pay for the solar project and led to deep energy-bill reductions for participating members of the local community. The resulting 2 MW solar array wouldn't have been possible in a state that didn't value solar more highly than polluting sources of energy or that didn't have additional incentives for investing in low- and moderate-income communities.

State solar policies define solar market economics, which in turn determine how large or small a parcel of land can be for siting a solar project that will pay for itself and generate savings for users and a return on investment for owners over the course of its useful life. That means you need to understand how the land that's available in your community lines up with the acreage and type of land that is needed for solar projects being developed in your state.

If you live in a state with strong solar policies, you'll be able to build solar projects on smaller parcels of land and still be able to share savings and other benefits with the local community, similarly to Rockford. If your state doesn't have a renewable portfolio standard or other renewable or solar policies in place, you'll need larger tracts of land to build big projects that are cost-competitive with big power plants based on scale, and it's less likely that smaller projects will pencil out.

Using the Sun and Land

Once you know how big or small a solar project your state's market can support, you need to understand how your community's land relates to the electricity transmission and distribution grid. Large or small, solar projects have to be connected to the existing grid, which may be old and congested or modern with plenty of room to accommodate new sources of power and better technology. While it's critically important to modernize and upgrade our electricity grid for the safety and security of our communities for the long term, solar projects on land that is open with a topography that is oriented toward the sun and that's close to modern, high-capacity transmission and distribution infrastructure costs the least to build and offers the greatest savings. That means you'll want to focus on right-sized, open, solar-oriented land near good electricity infrastructure while also identifying areas of your community that are good for solar but are in need of an infrastructure upgrade. These areas, which may also be more vulnerable to power outages, may be better candidates for resilience-focused projects, which we'll address in the next chapter.

Smaller solar projects, including distribution-scale solar, may be installed on rooftops, on canopies over parking lots, or on parcels of otherwise unused land located within the fabric of an existing neighborhood. The majority of existing solar capacity, however, is installed in large-scale solar projects on hundreds or even thousands of acres. These projects are ground-mounted, which means that the solar panels are affixed to a metal racking system that is attached to the ground. Many also have motorized tracking systems that enable the panels to follow the sun's movement throughout the course of the day, much the way sunflowers do, to maximize the amount of energy the array generates.

Whether you're installing a solar project on ten acres near town or on a thousand acres of grazing land in the county, there are important land use questions to ask and answer at the local level. Most municipal planning, zoning, and building codes don't include guidelines for solar projects, because solar is a relatively new technology in building industry terms. If planning and zoning rules apply in your community, you'll need to work with local officials to put together a game plan so that they have a way to review and approve solar projects. Although it's a dense urban environment, the District of Columbia offers a helpful example. The District implemented a bold solar policy, but had building and zoning codes that didn't address solar panels. Homeowners and local businesses looking to install solar on their rooftops, parking lots, and available land ran into a problem. Was rooftop solar like the addition of an additional story to a home? Was parking lot canopy solar like a garage? Was ground-mounted solar on an open parcel of land like a new building? The answer to each question was "not really," but there was no such thing as a solar array as defined in the building and zoning codes. In order to enable new solar installations, but to do so in a way that respected both property owners and local neighborhoods, the District used existing concepts such as penthouses, fences, and setback requirements for additions to existing facilities to create a permitting pathway for new solar projects. Their smart, highly collaborative, and practical process identified and quickly resolved challenges to make way for the District's clean energy future without compromising other planning and zoning priorities. If your community is facing similar questions, the US Department of Energy, the National Renewable Energy Lab, and professional associations including the Urban Land Institute all have planning resources available that can help.

Similarly, some farming communities have raised questions about the long-term impacts of putting agricultural land to work generating solar. There is, however, no need to compromise between energy and US food security. It would take less than 2 percent of America's 847,400,000 acres of farmland to power the whole country with solar energy, and that's before you consider the potential of co-locating solar farming with other crops for dual land use.[7]

Dual Land Use

If you want to put the benefits of solar to work for your community and also preserve the agricultural purpose of the land, there are a number of innovative approaches you can use depending on the local climate and crops. Jack's Solar Garden in Boulder County, Colorado, is a leading example of how solar and agriculture can work together. Informed by research from Dr. Greg Barron-Gafford's team at the University of Arizona on agrivoltaics, Jack's Solar Garden includes a 1.2 MW community solar installation that shares power with local neighbors and businesses. It's 3,200 solar panels (enough to power about three hundred homes) are mounted high enough above the ground to grow crops underneath them, and the array is ringed by pollinator plantings. Scientists at nearby NREL will be tracking the results,[8] informed by Barron-Gafford's research and other global scientific studies that have shown that the shade from solar panels can help many crops thrive by shielding them from high heat and evaporation that would otherwise hurt their yields. Similarly, the plantings help keep the panels cooler, increasing their productivity. In a world with increasingly extreme weather, including heat and drought, pairing solar installations with agriculture may even help preserve farming in places where it would otherwise be threatened by the effects of climate change. Jack's Solar Garden, for its part, will help light the way. It's already impacting Colorado land use policies, which were updated in 2018 to allow community solar installations on land designated for agriculture.

To make solar projects everywhere more integrated into the landscape and the local environment, a global community of scientists and advocates is working at the intersection of biogeography, agriculture, and energy. Pollinator-friendly solar is another approach their community is advancing. Solar projects typically have turf grass or gravel installed underneath ground-mounted panels. Gravel installations, in particular,

AGRIVOLTAICS

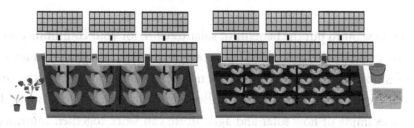

Agrivoltaics enable solar and food production to work together, for the benefit of both. (Original graphic.)

can take away from the beauty of the land, turning an agricultural area to a site that looks more industrial and potentially creating community opposition. Gravel can also cause a localized heat island effect that negatively impacts energy production by overheating the solar panels. In contrast, pollinator plantings honor the agricultural roots of the land—filtering stormwater, building topsoil, sequestering carbon, and supporting bees and other pollinators that are critical to farmers—while also eliminating gravel's heat island effect to keep solar panels producing at higher efficiency.

Rob Davis, the director of the Center for Pollinators in Energy with Fresh Energy out of Minnesota, is a leader in this global community. I first met Davis through the team at The Ray, a nonprofit proving ground for sustainable highway technologies that had succeeded in getting a 1 MW right-of-way pollinator-friendly solar demonstration project written into the 2016 Integrated Resource Plan approved by the Georgia PSC. At the time, the idea of pollinator-friendly solar was nascent, but emerging rapidly. Just one year later, in 2018, Minnesota's Xcel Energy would become the first utility in the United States to require developers to submit a pollinator plan as a part of their solar project. To date, seven states have followed Minnesota's lead, passing voluntary pollinator-friendly solar guidelines, and the State of Virginia has created a voluntary Pollinator Smart solar program to help developers apply best practices and identify the optimal native plants to use.[9] Though not yet widespread, pollinator-friendly solar—and agrivoltaics—will become household words, if Davis has his way. In March 2021, he launched a campaign to produce a pollinator-friendly solar Lego set that needs just ten thousand votes to go into broad production.

Land Ownership, Wealth, and Restoration

Farmers and farms can benefit from solar, supporting multigenerational wealth-building and keeping land in the family. But sometimes questions about who owns the land as it passes from generation to generation can rob families with rural roots of these opportunities. In 2016, I met Ed Brame, then chair of the board of a local church in the District of Columbia, while working on a community solar project to provide clean energy and utility bill savings to residents. Brame shared a story from his own family's history that stunned me.

In the late nineteenth century, like many African American families of the time, the Brame family migrated west and established a farm in Oklahoma. As generations passed, the title to the farm was not probated, which meant that multiple family members held a claim, and no one individual or group of family members held clear title. This made the Brame farm vulnerable to a predatory practice known as forced, or partition, sale—and that's exactly what happened. Another local landowner purchased a claim to title from one of the heirs and used that claim to force the sale of the property through the courts. While each of the heirs with a claim received compensation from the sale of the property, Ed's family lost the land that his family had owned for a hundred years.

Thomas W. Mitchell, a MacArthur Fellow and the nation's leading scholar on Black landownership in America, has written extensively about the role that partition sales have played in robbing Black families of land and multigenerational wealth. Black families are more vulnerable because they are much less likely to have registered a will than white families. While 65 percent of white families have wills, only about 23 percent of Black families do, and they are less likely to have access to legal services if their land ownership is challenged. Over the past hundred years, this practice, together with racially discriminatory lending and other actions by USDA, contributed to a massive loss of land and Black wealth. As of 1910, Black landowners held approximately seventeen million acres of agricultural land, but today that number is under two million acres.[10] Estimates of the value of lost land and opportunity range from $250 billion to $350 billion.[11] Forced partition sales and USDA's historic farm financing practices are both part of a long history of law and policy being used to deprive Black families of property. For example, Union general William T. Sherman's Special Field Order No. 15, which was crafted based on ideas proposed by a group of twenty abolitionist ministers led by Garrison Frazier, who himself had been enslaved until 1857 when he purchased his own freedom, implemented the "forty acres and a mule" policy to grant agricultural land to freed people to become self-sufficient and self-governing, and to build wealth. President Johnson, a Confederate sympathizer who was elected after President Lincoln, overturned the order and returned the land to the slaveowners who had seceded from and declared war on the United States.[12]

Thomas Mitchell's work includes a prescription for protecting landownership and preserving multigenerational wealth. Mitchell led the development of the Partition of Heirs Property Act, which is promulgated by the Uniform Law Commission and has since been adopted in seventeen states and the US Virgin Islands. The act provides due process protections, including a right of first refusal for other heirs, and requires that any sale be conducted on the open market. The 2018 Farm Bill includes incentives for states that have not yet adopted the act to do so, which is a critically important step to preserving multigenerational wealth and land ownership.[13] If your state has not yet adopted this important legal protection for multigenerational land ownership, now is a good time to ask your state legislator to get to work on it so that the expanded use of land resources for renewable energy also increases equity and justice.

The Solar Business Model: Connecting Value with Values

While policies, markets, and land use can be complex, the solar business model is straightforward for projects at every scale. From the perspective of the solar project itself, there are three sources of revenue: electricity sales from the energy the solar project generates, tax credits, and other incentives if they exist. On the expense side, there are construction and materials costs, including the cost of the solar panels; installation labor; utility interconnection costs and other taxes, fees, and permits; operations and maintenance; debt service if applicable; and solar lease payments for the project site.

The resulting product—solar electricity—may also provide savings and other types of benefits to consumers, including the ability to buy locally produced clean energy. There is also indirect and intangible solar value, including inspiring and educating young people for trades and careers in clean energy while showing them how you are investing in long-term economic and environmental health for their futures. Each value stream—revenue, expense, and product—can be connected to the local community. For example, local electricians and lineworkers can be hired to support project installation and solar savings can be shared with households with high energy burdens to improve local affordability and quality of life.

The Role of the Utility: Distributing Power and Value

Your local utility has a critical role in connecting the value of solar to the community because its infrastructure is how solar power is distributed and delivered. Whether your state has a regulated energy market monopoly or an open and competitive deregulated retail energy market, utilities are responsible for transmitting and distributing energy from its source of generation to the homes and business where the electricity is used. Even if you're using rooftop solar on your own home, your home remains "plugged in" to the grid with energy flowing in both directions.

The economic value of energy flows through and is distributed by the utility alongside power. In regulated state monopoly energy markets, utilities buy solar electricity and influence the rules by which new solar projects can connect to their grids, including the costs and fees associated with interconnection. Utilities can help or hurt. For example, Alabama's solar tax is a fee charged by the state's largest investor-owned utility for solar project interconnection that is so high that it prevents people from building solar on their own property for their own use. By contrast, in Washington, DC, Pepco Holdings has an entire team dedicated specifically to supporting utility interconnection of solar projects to streamline the process for its customers. Beyond the interconnection process, large investor-owned utilities, enabled through their regulatory processes, can also use their role in the market to form partnerships with communities that address place-based needs and opportunities. For example, Georgia Power built an innovative 1 MW solar project in Troup County on the Interstate 85 right-of-way in collaboration with The Ray to demonstrate how underutilized land can be used to generate energy. As a result, multiple states and the federal government have adopted new policies to encourage state departments of transportation to partner with utilities for renewable energy development and to explore underground direct current transmission.

Small-town public power and rural cooperative utilities have uniquely important roles in the distribution of solar value because of their community economic development missions. As locally owned and operated nonprofit utilities, they can partner with the community to develop solar projects and services in ways that meet the local community's unique needs. For example, Roanoke Electric Cooperative is using solar to advance economic equity. It partnered with EnerWealth Solutions, a Black-woman-owned solar and energy storage company founded by Ajulo Othow, to develop a community solar program that prioritizes working with small and minority landowners to help preserve multigenerational land ownership and build wealth. As the solar project developer, EnerWealth offers landowners a profit share of the revenues from the solar and energy storage installation located on their property in addition to lease payments. An additional profit share goes to a local community development nonprofit and EnerWealth's solar projects include pollinator plantings.

Developers and Investors: Values Alignment

Roanoke's collaboration with EnerWealth Solutions is a wonderful example of the benefits that can come from working with development and investment partners who share a community's values and priorities. Rockford Solar, the project developed on an abandoned rock quarry in Illinois, is another great example. Solar developers and investors who are solely focused on profits would be less likely to take the time or care necessary to understand the transformative potential of turning a quarry into a thriving solar farm or the importance of preserving Black landownership to help families thrive in their communities for the long term, because it's not their purpose or priority. It's an important consideration because solar developers and investors are essential to a successful project.

While the solar business model is straightforward, developing a solar project is highly technical and requires engineering, geotechnical, and financial expertise—particularly for large-scale solar projects that can cost tens of millions of dollars to build and hundreds of thousands of dollars just to get to a groundbreaking. Whether a developer is seeking out your community or you're looking to partner with a developer to bring a renewable energy project to your hometown, inquire about their values to make sure they align with the priorities of your community. When the development process and project economics are transparent and the developer seeks to build authentic relationships with the community, you've got two good indicators that you're talking with the right folks.

Values alignment is also important to seek among the solar investors who would own and operate the solar project in your community, and who may or may not be part of the same group that serves as developers. Similar to affordable housing developments, solar projects need investors who can use tax credits because a significant portion of the value of any solar project remains linked to federal tax incentives. In 2021, these incentives enabled investor-owners of solar projects to take 26 percent of the cost of the solar project as a tax credit. While the value of the tax credit is scheduled to step down over time from a high of 30 percent to just 10 percent for commercial solar investors-owners, it has important implications for who can realize the greatest economic benefits from solar. Because you have to have pay taxes to benefit from a tax credit and rural cooperative and public power utilities are not-for-profit, neither type of utility can use tax credits. That means that nonprofit utilities have to partner with values-aligned solar investors to get the full economic benefit of solar. Moreover, if owning the solar project is among the

community's goals, the nonprofit utility or other local nonprofit institution will need to use an innovative project finance structure to enable
ownership to flip back to the nonprofit after the tax credit period ends.
The National Rural Electric Cooperative Association's SUNDA (pro-
nounced "Sunday") Project offers many helpful resources for cooperatives, and Groundswell's LIFT Solar Everywhere research project offers
recommendations for other types of nonprofits.[14] Executing a tax equity
flip transaction can be technically complex, and you will need to make
sure you engage the support of an experienced lawyer who has worked
with tax credit transactions before.

Solar Project Types, Benefits, and Implementation Models

Now that you're equipped with the essentials of the solar business model
and understand how policy and your community's land use priorities
shape solar possibilities, you're ready to develop a solar strategy for
enhancing economic development, affordability, and local quality of life.
Before we look at specific implementation models and their benefits,
we'll introduce the three main types of solar projects: behind-the-meter
(or rooftop), community, and utility-scale solar. The solar technologies
are the same in each, but the scale of implementation and manner in
which each is connected to the grid vary in ways that impact project
economics, development strategies, and potential community benefits.

Types of Solar Projects

Behind-the-meter or rooftop solar is installed on homes and businesses
to support each building's own electricity use. While the direct benefits
of behind-the-meter solar primarily serve the resident or business that
owns the solar panels, there can be additional benefits for the community as a whole if residents and business join together to adopt solar.
Solarize programs, for example, organize groups of home and business

owners into a group purchasing commitment to negotiate better solar pricing through volume, thereby bringing down the cost of solar for everyone. Communities working together to adopt solar can also have public health and jobs benefits. The Hozho Homes Program from Native Renewables installs off-grid solar systems for Hopi and Navajo families, enabling them to own their own power while providing training for Native American solar installers. Because many Native American families in remote rural areas aren't served with electricity and have been forced to depend on diesel generators that produce pollution linked to respiratory disease, community-wide solar adoption also promises public health benefits.

Community solar enables multiple local households and businesses to share solar power from a single nearby solar project, which might be located next door, down the block, or across town. Echoing the original formation of rural cooperative and small-town public power utilities, this type of solar project can pool the buying power of local residents and businesses to support clean energy projects that meet the community's needs. A community solar project might be installed on a local school or church, or in a vacant lot, but its capacity is typically smaller than 5 MW. Among the fastest-growing types of solar projects, community solar expands solar access for people who rent their homes, don't have the financial resources to purchase a solar system, or prefer not to install solar on their roof. It can also have important infrastructure benefits because community solar enables power generation at the distribution scale, which may help utilities reduce costs, improve efficiency, and improve resilience.

Covering hundreds or even thousands of acres of land, utility-scale solar projects can generate enough electricity to power large data centers, commercial and industrial facilities, or even whole communities. These large solar projects are developed to meet the energy demands of utilities or corporate energy buyers. The benefits of utility-scale solar include the economic value of multiyear land leases for the property where the solar project is located, construction and engineering jobs, and affordable electric power that is less expensive to produce than energy from other sources, even without incentives.

All three types of solar projects can be combined as a part of local economic and community development strategies to attract new investments, expand community benefits, share power, and develop new energy services that expand access to affordable, clean power. The best approach to take will depend on what your community needs and what it has to offer.

Attracting Investments and Jobs

Being able to provide local solar power is a competitive advantage for attracting investment and jobs because an increasing number of companies will build new facilities only where they can purchase affordable renewable energy. To date, more than three hundred corporations have committed publicly to using 100 percent renewable energy in their operations. As a measure of the scale of the corporate renewable energy market, global corporate purchases hit a record of 23.7 gigawatts in 2020, with the United States leading the way. Amazon.com alone purchased enough energy to power 1.53 million homes.[15] The impact of large corporate renewable energy purchases extends far beyond the power sector. The Clean Energy Buyers Association (CEBA), the membership organization for large corporations who are working together "to create a resilient, zero-carbon energy system," boasts more than 230 members with a combined $5.8 trillion in annual revenues and more than 13.5 million employees in the United States alone.[16]

Corporate commitments to 100 percent renewable energy enable you to put your local solar capacity, and your local utility's solar capabilities, to work as an economic development strategy to attract new businesses to your hometown. Conversely, if your local community can't provide renewable energy for corporate customers, it won't be able to compete. As a result, over time, the lack of competitiveness will lead to job loss, revenue loss, and a further loss of community vitality as local budgets shrink.

That's exactly what was happening among Georgia's public power utilities prior to 2019, because they couldn't sign the long-term contracts for renewable energy that corporate buyers needed. At the time, Georgia law set a ten-year limit on the term of local government contracts, which also applied to public power utilities as units of local government. The typical corporate renewable energy contract, however, ranged from fifteen to twenty years, the length of which was necessary for securing project financing. When large electricity customers that demanded renewable energy put out requests for proposals to select a utility provider, small public power utilities couldn't meet their contract terms and therefore couldn't compete. For example, the City of Covington's public power utility got left out when Facebook decided to build a new data center in Newton County, in which the City of Covington is located. Walton Electric Membership Corporation, a rural electric cooperative utility with extensive renewable energy experience, won Facebook's business.

I experienced the impact of those constraints firsthand through Groundswell's work in LaGrange. While the local community's top priority was reducing energy burdens for residents with low household incomes, there was also widespread interest in solar. The nearby Georgia International Business Park was already host to a number of early corporate sustainability innovators, and Economic Development Authority president Scott Malone viewed demonstrating the community's ability to deliver renewable energy as a competitive priority for attracting new business. Working with our partners at The Ray, the Groundswell team had connected with local landowner and real estate entrepreneur

Jim Daniel, whose family still held his father's four-hundred-acre farm, which was situated in the middle of the Georgia International Business Park near the kind of utility infrastructure needed to interconnect a large solar project.

The project's vision began modestly, with about fifty acres of community-based solar anchored by Walmart's nearby regional distribution center. However, the project expanded rapidly when Walmart expressed an interest in purchasing enough renewable energy to meet the needs of all its facilities served by public power utilities across the state, a step toward achieving its corporate goal of 100 percent renewable energy by 2035. While the project held great potential to connect the community's vision with Walmart's energy and climate goals, it quickly hit Georgia's ten-year contracting wall. Rather than giving up, the team worked together to change the law—enabled by Walmart's commitment to small towns and renewable energy, supported by the City of LaGrange, and catalyzed by the leadership of The Ray. Senate Bill 95 was authored and sponsored by Georgia state senator Randy Robertson, and was introduced, unanimously passed, and signed by Governor Kemp during the 2019 Georgia legislative session. As Senator Robertson has explained, the rationale for the legislation was "to help bring a level playing field to our Electric Cities . . . and to create additional and cleaner energy choices, all leading to better quality of life for all Georgia families."[17] The new law enabled municipal utilities to enter into long-term contracts of up to twenty years for solar and wind projects, which was a powerfully important economic advancement for small-town Georgia. As a result, later that same year, the Municipal Electric Authority of Georgia issued its first-ever request for proposals to purchase 100 MW of solar generating capacity to help meet corporate demand, and by doing so created a process by which small-town public power utilities could compete for and win new corporate renewable energy business like Walmart's.

Sharing Community Benefits

Attracting new investments and jobs isn't the only potential benefit of corporate renewable energy purchases. Relevant initiatives are still relatively new with few case studies to review, but it is possible to deliver other kinds of community benefits alongside large-scale solar power. Community benefits can encompass supporting local businesses, training, and jobs; sharing savings; preserving habitats; sharing resilience benefits; restoring communities; preserving multigenerational farm ownership; and more.

A handful of leading companies that have committed to purchasing 100 percent renewable energy have also been innovators in the development of solar projects that demonstrate how to deliver more than renewable energy. Google, for example, took a pioneering approach and converted a retired TVA coal plant in Alabama into a data center powered by 100 percent renewable energy that helped keep local people employed.[18] Announced in 2015, the project was Alabama's first large-scale data center project. In 2019, Salesforce shared its experiences in a white paper titled "More Than a Megawatt." The paper begins with the observation that not all renewable energy projects are created equal—even if the economic details of the transaction are identical. As the authors noted, "some renewable energy projects displace more fossil fuels than others, some are built at the cost of critical habitat for plants and animals, and others provide invaluable support for their local community."[19]

Building on these early innovations and responding to surging corporate interest in how to use renewable energy investments to deliver additional community benefits, Groundswell launched a research project with CEBA and the Solutions Project to create a road map that more companies could follow. Central to the effort was the four-part Working Wisdom Listening Tour, which focused on learning from real-world

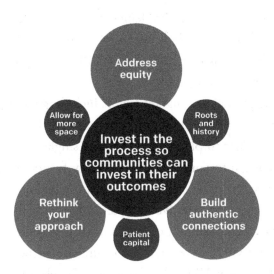

Strategies for connecting corporate renewable energy purchases to local community benefits and priorities. (Source: Groundswell; original graphic.)

projects and community-based energy innovators to inform the project. As a result, detailed guidance for corporate renewable energy buyers and a procurement-focused decision support tool to guide the process are now available online.[20] While these represent first steps, and additional tools and strategies are still being developed, implementing their recommendations can have broadly beneficial results by connecting highly technical and distant procurement processes with the needs and priorities of local communities in innovative ways.

Sharing Solar Power

Community solar projects are unique in their ability to help neighbors within the same utility service territory share power from a solar project. Rural electric cooperatives, in particular, have long been leaders in community solar. The National Rural Electric Cooperative Association (NRECA) even published a playbook in 2016 to help local utilities develop business models and implementation plans for bringing community solar to their member-owners, though ironically, at the same time the playbook was being developed, NRECA was opposing clean energy legislation in Congress.[21]

Because energy policies, market factors including energy pricing and price structures, natural and built infrastructure, and community priorities vary from place to place, there's no one single way to incorporate community solar into your local clean energy future. There are, however, many great examples to look to for inspiration and guidance.

Vernon Electric Cooperative completed the very first community solar project in Wisconsin in 2014, and it has been sold out ever since. Local member-owners were invited to purchase shares of the 1,001-panel array located at VEC's headquarters. The electricity production from each member-owner's panels are credited against their utility bill, and panels can be bought and sold among members.[22] Its model has created a strong sense of community ownership among participating

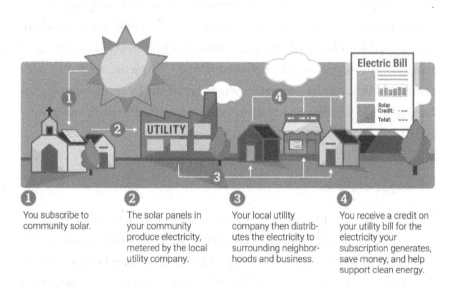

1	2	3	4
You subscribe to community solar.	The solar panels in your community produce electricity, metered by the local utility company.	Your local utility company then distributes the electricity to surrounding neighborhoods and business.	You receive a credit on your utility bill for the electricity your subscription generates, save money, and help support clean energy.

How community solar works. (Source: Groundswell; original graphic.)

residents. When heavy snows blanket the solar farm, members regularly call into the utility to make sure their panels are being cleared to keep them producing power.

Walton EMC, the same rural utility that won Facebook's clean energy business, built its first 1 MW community solar project in 2015, at the urging of its CEO's daughter. Member-owners can purchase up to two blocks of community solar at twenty-five dollars per block, the production from which is credited to the member-owner's utility bill.[23] Located on six acres next to its headquarters in Monroe, Walton EMC's first community solar project sold out, and the program quickly expanded to two additional sites. This early community solar leadership positioned Walton EMC to become one of the Southeast's leading utilities in solar development, attracting new businesses and investment to its community.

Building on this history of community solar leadership, the US Department of Energy awarded a $1 million multiyear research award to NRECA to study pathways for making solar more accessible and affordable for rural households with low and moderate incomes. The resulting ACCESS project, which stands for Achieving Cooperative Community Equitable Solar Sources, was led by Adaora Ifebigh, former NRECA program director for Energy Access. The body of knowledge the ACCESS project is creating is a rich resource for understanding how rural cooperative utilities can use community solar to meet their unique, place-based needs.

The Anza Electric Cooperative's SunAnza and Santa Rosa Solar projects are among those being studied. The Anza Electric Cooperative was born in the 1950s, when local residents, frustrated with the high cost of power from California Electric, reached out to the Rural Electrification Administration and began the process of funding and forming their own local electric cooperative.[24] Located in rural southwestern California, Anza serves 5,200 residential member-owners and has a long-term agreement to provide power for the Santa Rosa Band of Cahuilla Indians. The success of its first solar project, SunAnza, demonstrated how solar could help reduce peak electricity costs for the utility and its members. As a result, Anza began work on a second community solar project on tribal lands. By pairing community solar with a time-of-use tariff that charges less for electricity usage at off-peak hours, the Santa Rosa Solar project will share annual savings estimated at $600 to $1,000 per year per participating household.[25]

Always among the vanguard, Roanoke Electric Cooperative takes a different approach that puts philanthropic support to work, leveraging other programs to deliver extra savings to member-owners through the SolarShare program. Philanthropic donations are used to pay for no-cost community solar shares for low-income members-owners, which in turn reduce their utility bills. The first fruits of the resulting savings are used to cover the costs of home repairs, such as fixing leaky roofs, that are necessary to enable energy upgrades such as insultation that deliver even more savings through energy efficiency.

Rooftop Solar as a Service

Cooperative and public power utilities are also pioneering programs that enable member-owners and customers to install rooftop solar systems. Instead of people having to purchase and install solar systems on their homes on their own, these utilities are showing how rooftop solar can be offered as an energy service that provides savings to the customer and benefits to the community. It's a critically important demonstration, because if everyone who could afford to buy their own solar system decided to go it alone without considering how they might work with their neighbors, people who didn't own their roof or couldn't afford to buy a solar system would be left out of many clean energy opportunities.

Making residential rooftop solar a service provided through the local utility keeps the public in public power.

One organization making this a reality is CPS Energy, the public power utility that serves San Antonio, Texas. One of the largest public power utilities in the country, it's also among its most innovative. CPS worked with PowerFin Partners to launch the SolarHost program, which enabled CPS customers to host a rooftop solar system with no installation costs while also sharing in the solar savings. Through the program, participating CPS customers allowed PowerFin to install a solar system on their roof. In return, the participating customer received utility bill credits of $0.03 per kilowatt-hour for the energy produced by the panels over a term of twenty years. Announced in 2015 and currently fully subscribed,[26] SolarHost sets a great example that other utilities can follow. By working with a third-party partner that can benefit from the solar tax credit, and structuring the program so that residents "host" solar panels in return for savings rather than purchasing a system themselves that they then have to maintain, the SolarHost program sets a creative example for how to share the benefits of clean energy installed on individual homes.

Ouachita Electric Cooperative is another early pioneer that's expanding access to residential rooftop solar. Through its HELP PAYS energy efficiency program, Ouachita's member-owners can finance up to 80 percent of their annual solar savings.[27] As research released through the US DOE–funded LIFT Solar Everywhere project has found, Solar PAYS would be much more broadly applicable as a tool for rural cooperative and small-town public power utilities for expanding affordable solar access if Congress took action to enable nonprofit utilities to convert the solar tax credit to a grant. Until then, local nonprofit utilities will need a values-aligned investment partner to realize the full economic benefit of solar.

Abundance

The opportunities for supporting your community's vitality with solar power are abundant. You can begin where you are, work with your state's energy policy and market structure, change the rules if you need to, use the land well, work with partners seeking to do more and do better based on shared values, and choose from among a variety of solar scales and models to match up with your and your neighbors' needs. It's liberating compared with the constraints of a scarce fossil fuel–powered energy system that had to take resources from one place and burn them in another to create enough power to support new industries and opportunities.

As you're building your solar strategy, don't fall back into old ways of doing things when you have the chance to build a new system based on better values—a system whose benefits can be seen within your generation. Remember the reparative and restorative approach to solar development pioneered by EnerWealth Solutions in North Carolina, and always keep in mind why rural cooperatives and public power utilities were created in the first place: to enable rural and small-town communities to come together to own their own power. The field is wide open and the harvest is ripe. Take deep breath, roll up your sleeves, and get ready for the joyful work.

CHAPTER 7

Energy Resilience
and Self-Reliance

Resilience is a powerful word. It can be applied to any number of natural, built, and social systems and refers to the ability to recover function following a significant, potentially unpredictable disruption. As it relates to our energy systems, moving away from long transmission lines and centralized power plants that burn polluting fuels and toward distributed systems that combine local energy storage with abundant renewables improves resilience. It's another way that technology is becoming better aligned with our environment and the distributed nature of how we make decisions about energy.

New, clean, resilient, and renewable energy technologies couldn't be emerging at a better time. Most of America's power lines were strung in the 1950s and '60s with a predicted lifespan of about fifty years, and some of our grid infrastructure dates back to the late nineteenth century, so it's well past time for a major investment in modernization. Our aging grid infrastructure is among the reasons that the United States experiences more power outages that any other developed country in the world.[1]

Old technology also makes our grid more vulnerable to emerging sources of disruption, such as cyberattacks, which are on the rise. For example, in 2015 in Michigan, the Lansing Board of Water & Light paid hackers a $25,000 Bitcoin ransom to regain access to its communications systems after being locked out by a cyberattack.[2] In a more recent example, in 2021, the Colonial Pipeline Company paid a multi-million-dollar ransom to hackers who took down the largest fuel pipeline in the nation, driving gas prices up and leading to local fuel shortages.[3]

The urgent need to modernize our grid extends beyond aging physical infrastructure and vulnerable technology. Climate change is making our weather more intense and less predictable, which is inexorably linked to our energy systems because pollution from the fossil fuel energy we've depended on since the industrial revolution is causing our climate to change. On our current path, weather that would once have been considered extreme is becoming the norm. These changes in our climate, caused by what kind of energy we've used for the past five generations, mean that we must make changes during our lifetimes too. Moving quickly toward zero-carbon forms of energy is only part of the picture. Building for resilience acknowledges that our future isn't going to be like our past, and we therefore need to plan and build differently to meet its challenges.

But resilience isn't just about making it easier to bounce back when bad things happen, it's also about creating new opportunities for economic development and future-focused growth. The same systems that keep the lights and heat on when the grid goes down can also reduce infrastructure costs, cut electricity bills, and create new energy services that cooperative and small-town utilities can offer their customers to help increase revenues to support community development and beneficial growth.

Energy resilience builds on each of the clean energy solutions we've explored together already. Incorporating technologies that build resilience, like connecting local solar generation and energy storage through nimble microgrids, expands the range of benefits each clean energy solution can offer while accelerating the localization of our energy system.

Pliny Fisk, cofounder and codirector of the Center for Maximum Potential Building Systems in Austin, has observed that the boundary between the forest and the field is where all the most interesting things happen. Energy resilience is right at that innovative edge of our transition from an industrial energy past to clean energy futures. It's exciting and will take a commitment to innovation to incorporate resilience into our local clean energy futures.

In that spirit, in this chapter, we'll explore how resilience technologies, benefits, and business models are coming together to create opportunities for increasing the energy resilience of your community. As we do, keep in mind that the same state and local policy factors that impact the solar market will likewise impact your approach to building energy resilience. Because policies differ from place to place, the optimal approach to energy resilience will also vary from place to place. For these reasons, we'll review examples of the many ways communities are building energy resilience and hopefully inspire ideas about where you might begin. But first, we'll revisit the experiences of Texas during the 2021 power crisis, California during the ongoing drought and wildfire crisis in the American West, and New York during Superstorm Sandy as indicators of how important it is to address energy resilience in the context of climate change.

Localizing Energy Systems

During February 2021, an estimated fifteen million Texans went without power and hundreds of people died during extreme cold weather that saw every single Texas county under a winter storm warning at the same time for the first time in recorded history.[4] While it can get cold in Texas, extreme cold hitting the entire state all at once for an extended period was unprecedented, and it overwhelmed the state's electric power system, froze natural gas wells and pipelines, and left residents that depend on electricity for heat without power for days in temperatures at or below zero. It's no surprise that interest in residential solar and energy storage systems surged following the crisis as people sought to take keeping the lights on into their own hands.[5]

California is facing a different set of climate crises, including the most expansive and long-lasting western drought in a century, which has contributed to hundreds of thousands of acres of unrelenting wildfires; the loss of life, habitat, and property; and a fire season that seems never-ending. Wildfire conditions also result in power outages when utilities shut down the grid to prevent their equipment from sparking fires. As a leading indicator of the future impact of climate change on the utility sector itself, Pacific Gas & Electric, which was the first utility to commit to the Paris Climate Agreement, also became the sector's first climate change casualty when the company declared bankruptcy as a result of more than $30 billion in legal claims over the role of its overhead lines and equipment in sparking wildfires.[6]

Water, power, and fires in California are deeply intertwined. About 20 percent of California's electric power is used to treat, heat, and pump water, and about 11 percent of the state's electricity generation comes from hydropower. Water shortages due to the drought also mean less hydropower, which leads to rolling power outages in high heat, when people need more electricity to keep cool, adding yet another layer to the California energy sector's climate crisis.[7]

Extreme cold, extreme heat, and extreme drought are all affecting communities across the United States, and so is extreme flooding. Back in 2012, Superstorm Sandy hit every state on the Atlantic Seaboard, from Florida to Maine. In New York City, storm surges of corrosive salt water flooded tunnels that housed transmission lines, underground substations, and other energy infrastructure. Because restoring flooded electrical equipment is harder and takes longer than repairing downed overhead transmission lines, restoring power in some parts of the New York City metro area took weeks.[8]

Each state in these three examples has different vulnerabilities that need to be understood to develop the right energy resilience strategies to meet the needs of its residents, and the same is true of your community. For example, if you live in the Pacific Northwest and never needed air-conditioning before, but now you're experiencing longer stretches over 100 degrees, your community might consider prioritizing solar and energy storage for senior centers and schools to help your most vulnerable neighbors keep cool during rolling power outages. The challenges will be a little different everywhere, so it's important to take the time to confront how climate change will impact the place where you live with the same courage and spirit with which you embrace innovation. There is a large and growing volume of resources available to help you navigate this complex topic, including in the realm of climate science and its applications. The National Oceanic and Atmospheric Administration, the same agency that tracks weather and hurricanes, publishes a series of helpful maps that show where drought, extreme heat, and extreme weather are occurring. Available at Climate.gov, these resources are an informative place to see how climate change is impacting your state and region. Building on the data, the American Planning Association offers a wide range of planning resources, from model policies to strategic frameworks, that can help you apply the science to your work.

Resilience Technologies: Microgrids and Energy Storage Hardware and Software

As you're getting to know your community's individual energy resilience needs by familiarizing yourself with the flood, fire, extreme weather, and other risks you face from climate change, the next step is to understand the building blocks of energy resilience, including microgrids and energy storage hardware and software. Keep in mind, resilience isn't just about mitigating risk; it's also about opportunities like increasing efficiency, reducing infrastructure costs, and developing new energy services—so energy resilience can bring many benefits to your community.

While you need to understand what's on the resilience menu, you don't need to know the recipe, so don't feel like you must become a climate science or energy technology expert to develop a strategy that works for your community. Building a resilient clean energy future takes a team, and you will find people who share your vision and can contribute to it, and who have the technical skills and capabilities needed to realize that goal, as you undertake your resilience research and planning process.

Microgrids: Local, Connected Energy Neighborhoods

Microgrids are like local energy neighborhoods, while our national energy grid is more like the interstate highway system. From a more technical perspective, according to the National Renewable Energy Laboratory, "a microgrid is a group of interconnected loads and distributed energy resources that acts as a single controllable entity with respect to the grid."[9] Microgrids can continue to operate and deliver power to the buildings and facilities they serve whether they're connected to the grid or not, but adopting microgrids doesn't require tossing our existing electricity transmission and distribution grid out the window. Microgrids can be integrated with the existing grid, making the entire system smarter, more efficient, and more resilient.

There are three key attributes of a microgrid, according to industry publication *Microgrid Knowledge*. Microgrids are fundamentally local and serve nearby buildings, homes, EV charging stations, and other facilities that use energy with electricity that's produced locally. Microgrids are generally more efficient than the centralized power plants and

MICROGRIDS

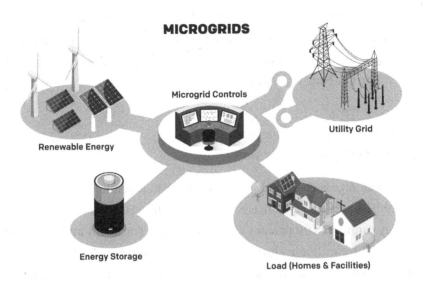

Microgrids can connect energy generation, storage, homes, and buildings to improve efficiency and resilience. (Original graphic.)

long transmission lines that make up our current energy grid because, depending on which state you live in, anywhere from about 2 percent to more than 13 percent of the electricity that's generated gets lost in transmission.[10] In addition, microgrids can operate independently in what is often called islanded mode. Even if the microgrid is connected to and operating in concert with the larger grid most of the time, in the event of an outage, cybersecurity breach, or other disruption, the microgrid can disconnect and operate on its own to keep the power on. Finally, microgrids are operated by a centralized software control system that manages energy use and storage within the local microgrid to optimize the function of the system as a whole.[11] Advanced microgrid control systems, for example, can maximize efficiency and minimize energy costs by managing energy time of use down to the appliance level. For example, an advanced microgrid might know to ease up on the air-conditioning in the middle of the day when you're not home and instead use power from your neighborhood community solar array to charge the battery in your storage system so you can charge your EV when you get back from dinner.

Energy Storage: Hardware and Software to Keep the Lights On, and More

Energy storage is an important complement to renewable energy because it resolves two major concerns about whether 100 percent renewable energy can meet 100 percent of our needs: intermittency and the "duck curve." Intermittency simply refers to the fact that the wind doesn't always blow nor does the sun shine all the time, and the duck curve is a graph that shows the difference between solar production and electricity demand over the course of a day. Without energy storage, the intermittency of renewable energy resources and the duck curve–shaped mismatch between solar production and energy demand would make it impossible to meet all our energy needs from renewables alone. Energy storage enables us to bring renewable energy production and electricity demand in line. Moreover, it enables us to store energy locally so that we have energy access to keep the lights on even if there are outages on the grid due to storms, fires, extreme weather, cybersecurity breaches, or other disruptive events. Linked through a microgrid, individual homes and buildings with solar and storage can be connected to one another to create a mutually supportive community that benefits everyone.

Energy storage systems are made up of hardware that stores energy and software that controls it. Batteries, which use electrochemistry, are perhaps the most familiar type of energy storage hardware. The most common battery technologies are lithium-ion, used in cell phones, computers, and many electric vehicles; and lead acid, used in many everyday applications (this is the type of battery that powers the ignition system in a gas-powered car). In addition, many promising new electrochemical and nanotechnologies are being developed by scientists and engineers to increase the efficiency and useful life of batteries and to improve their sustainability. Beyond electrochemical batteries, other types of energy storage technologies that can be applied at different scales and in different types of utility and industrial applications include thermal storage such as liquid salt, mechanical storage such as compressed air, pumped hydro for very-large-scale energy storage, and hydrogen technologies.

While there are many options for energy storage, the most available and best match for your home, building, or neighborhood given current technologies is probably the lithium-ion battery. Lithium-ion batteries are relatively affordable, accessible, and compact. For example, a battery storage system for a local church or school might be about the same size as the building's air-conditioning unit, so installing energy storage at your home might be like installing a new appliance.

Energy storage hardware is important, and we can look forward to science leading us forward with great strides in efficiency, affordability, and sustainability; but even the most cutting-edge hardware is of little use without the right software. Software is at the heart of energy storage's value proposition, including how it generates revenue for battery companies, and savings and other benefits for their customers. From controlling whether a battery or other device stores or dispatches energy to tracking real-time energy pricing and maximizing economic benefits to monitoring the usage and efficiency of the device itself—it's all in the software. For these reasons, smart software is as important as choosing the right type of energy storage hardware for your project.

While the electrification of the transportation sector and all the opportunities it presents is a topic unto itself that we'll explore in the next chapter, it's worth mentioning here that electric vehicles that are connected to local microgrids through bidirectional charging infrastructure can also provide energy storage and support resilience and its many benefits. In a charming recent example of how electrifying transportation improves resilience, a 2021 Ford F150 hybrid pickup truck saved a backyard wedding in Farmington Hills, Michigan. As the *Detroit Free Press* reported, when the power went out, wedding guests were able to plug the lights and the sound system into the truck itself and celebrate on into the night.[12] If one hybrid electric pickup truck can save a wedding, imagine what a network of local solar projects paired with batteries can do for your community.

Benefits of Energy Resilience at Different Scales

Now that you have a familiarity with how climate change is increasing energy vulnerabilities and a broad sense of resilience technology options, you can examine the scale at which you might begin introducing resilience into your clean energy future. Similar to solar, there are behind-the-meter, community- or distribution-scale, and utility- or grid-scale approaches. Each approach has different options for project funding and financing that line up with the value it generates.

Behind-the-meter energy resilience systems are building-scale solutions for homes, apartments, and commercial and industrial facilities that store energy for use by the building on which the system is installed. In residential applications, you might install an energy storage system paired with solar panels to reduce energy costs and make sure your home would have power if there were an outage. If your neighborhood had a microgrid and incorporated smart appliances and advanced energy efficiency measures like Alabama Power's Smart Neighborhood project at the Reynold's Landing in Hoover,[13] you'd enjoy additional benefits by enabling your home's energy systems to work in tandem with other home energy systems, reducing costs and improving comfort and affordability for the whole neighborhood.

Commercial and industrial facilities with behind-the-meter solar and energy storage have similar benefits, though they may be used and valued differently. For example, hospitals and data centers can't afford to have any downtime due to power outages and may have specific mission-related goals to operate on 100 percent clean energy 100 percent of the time. Incorporating energy resilience at the building level maintains continuity of operations, minimizes any risks from outages, and helps to match the facility's energy usage with on-site energy production to help meet corporate 24-7 renewable energy goals. For community-centered organizations such as churches and other communities of faith, resilience can also help extend the mission. For example, Empowerment Temple, a historic AME Church community in Baltimore, is installing solar and energy storage to serve as a community resilience hub for the surrounding neighborhood as a part of a nation-leading City of Baltimore program. When major storms roll through and the power goes out, neighbors can find a place to stay warm or cool and safe in the church's sanctuary.[14]

Community partnerships with large commercial and industrial energy users that value resilience for risk reduction can also create mutually beneficial opportunities for sharing resilience. John Kliem, formerly the head of the US Navy's Resilient Energy Program Office, oversaw a number of early examples that leverage the military's resilient energy leadership. An award-winning US Navy solar and energy storage facility developed in collaboration with the Kaua'i Island Utility Cooperative improves resilience for a local military facility while supporting the cooperative's goals. Kliem, who is now a business leader with Johnson Controls, has also suggested that co-locating energy storage with critical infrastructure such as hospitals and municipal water facilities could provide opportunities for sharing benefits like clean water and life-saving power.

In most parts of the United States, homes and buildings are required to be connected to the utility for public health and safety reasons as a part of the building code, but in remote areas, solar combined with energy storage can make you not only more resilient but self-reliant besides. Without access to solar or other sources of distributed renewable energy, remote communities that aren't connected to an existing electricity grid are dependent on dirty diesel generators for power. Not only is that an expensive way to keep the lights on, but pollution from diesel generators contributes to childhood asthma and other chronic respiratory disease. For example, more than fourteen thousand homes in the Navajo Nation don't have access to electricity, though ironically Navajo lands have been extensively exploited for extracting uranium, coal, and oil for energy.[15] Organizations like Native Renewables are bringing power to remote Native communities through off-grid solar and energy storage systems that are electrifying homes for the first time.

Community- or distribution-scale energy resilience moves from the individual building to the neighborhood scale, and can be a mosaic of individual homes, apartments, and other facilities with building-level solar or storage connected on a microgrid; or community solar with paired energy storage connected to local homes and buildings on a microgrid; or a combination of all the above (and more). Using the intelligence of microgrids to create a resilient energy community enables people and businesses to share the benefits. For example, Groundswell worked in service to Spelman College, Morehouse College, and the Atlanta University Center in collaboration with Partnership for Southern Equity and Georgia Power to develop the design for a new resilience hub serving the community. Enabled by funding from the National Renewable Energy Lab through the Solar Energy Innovation Network, when constructed, the project will combine behind-the-meter solar for the campus and a community center with energy storage on a microgrid. Notably, the project design process included residents from the surrounding neighborhoods to create a shared vision of how a resilience hub could benefit the community as a whole. In the case of an outage, stored energy will be sufficient to maintain critical electrical loads, including lighting and cooling for shared spaces where neighbors could come to keep themselves and critical medicines cool. According to the National Weather Service, heat has been the number one weather-related cause of fatalities nationally over the past thirty years, and elderly residents are particularly vulnerable to it.[16]

Energy resilience—safe, reliable electricity—is the core business of utilities. Increasing energy resilience at the level of utility operations is the product of combining a host of solutions at multiple scales, which may range from building- and neighborhood-level systems to massive pumped storage hydropower. Whether it's big renewable projects with big energy storage solutions or multiple smaller distribution-scale solar projects paired with storage, it's essential for local utilities to put their expertise and operational scale to work building resilience because they are in the best position to realize the resulting cost benefits. For example, Roanoke Electric Cooperative and EnerWealth are deploying community solar projects with energy storage that will increase local resilience by locating clean energy generation and storage within the community. This approach also creates cost savings for the cooperative by using stored energy resources at peak hours, thereby reducing the amount of peak power they need.

Large investor-owned utilities may also reduce infrastructure costs by investing in local energy resilience through distributed renewables, energy storage, and microgrids. In areas where the grid is congested and can't add any further electrical load without significant investments in transmission and distribution infrastructure, building local solar and energy storage may be a cost-effective way to relieve the existing grid's limitations. Where communities are focused on attracting new businesses that use more power, resilience can be an important strategy for supporting new jobs and economic growth.[17]

Funding and Financing Energy Resilience

Energy resilience holds tremendous promise, but how do you pay for it? Like solar and other renewables, the business models for energy resilience, including microgrids and energy storage, have to work within your state's energy market and regulatory structure, and also be compatible with the type of utility that serves your community. Within this context, make sure you have a clear picture of who will own the assets of your community's energy resilience project, because the best ownership structures will vary from state to state and by local utility type. Cooperative and public power utilities, because they are nonprofits, will need to partner with a tax equity investor to fully realize the economic benefits of tax incentives. In other scenarios, it may be optimal for different partners, potentially including the local utility, tax equity investors, and members of the community, to own different parts of the system and work as a collaborative to operate it. For example, a tax equity investor might own distributed solar assets to get the most out of the tax incentives, while the local utility might own the microgrid, and community members might own their own in-home energy storage appliances.

Mapping the value of resilience—which includes but is not limited to its economic value—to the funding and financing strategies you use to organize the capital needed to develop and build your first resilience project is a helpful way to cut through the complexity. Some of the value streams associated with energy resilience include innovation, reducing risks, optimizing renewable energy goals, reducing infrastructure costs, and earning revenue by selling electricity and providing demand response services. Identifying which of these approaches apply to your project will help you build your team and secure the stack of funding, financing, and revenue sources that will enable you to build it.

Because energy resilience is delivered through cutting-edge technologies that are in the early stages of commercialization, institutions including the US Department of Energy and many state agencies are investing in project predevelopment, demonstration projects, and research-focused activities. These investments, often structured as grants and technical assistance, place a value on innovation and learning and help bring down the cost of being among the first resilience projects in your state or locality. For example, the Maryland Energy Administration's Resilient Maryland Program has offered grant funding over multiple years to solar and energy storage projects that demonstrate results and share data, thereby helping to expand the market for resilience in Maryland. These grant funds have not only supported multiple completed resilience hubs but also helped illuminate the obstacles small-scale energy storage projects face when they seek to sell power on the wholesale electricity market.

Reducing risk is a feature of resilient energy with a quantifiable value that some companies pay for, which creates a potential revenue stream. For example, downtime associated with even occasional power outages may be very costly for manufacturers. Outages not only slow production but may also cost a great deal because it's difficult to bring some types of manufacturing processes back online when the power goes off. Companies that face these kinds of risks might previously have invested in backup diesel generators to prevent downtime. Solar combined with energy storage on a microgrid that prioritizes critical electrical load is a cleaner and more efficient solution. Small-town public power and rural cooperative utilities are beginning to offer energy resilience as a service to these kinds of corporate and institutional customers—medical facilities and technology and data centers, for example. New services for small nonprofit utilities mean revenue growth from innovation instead of from selling more or more expensive electricity; that means more revenue that can be invested in community development.

Some large renewable energy buyers and communities have set bold goals to operate on 100 percent renewable energy 100 percent of the time. Because renewables including wind and solar are intermittent, meeting these goals requires solutions that match renewable generation with demand. Because resilient energy systems also address intermittence and the duck curve, mentioned earlier in this chapter, they can be part of the solution set to help big customers and local communities meet their energy goals. For local utilities, offering these services can also be part of their contribution to local community economic development by attracting new corporate facilities through offering advanced energy service capabilities.

Reducing infrastructure costs can be a rich value stream for local utilities investing in energy resilience. For small utilities that have to pay more for wholesale electricity during peak hours of usage, pairing local distributed solar with energy storage can improve resilience and reduce the cost of electricity—which also benefits local members and customers. The scope of infrastructure cost benefits may even extend to other types of related utility services, including clean water and water treatment. For example, investing in resilient energy systems to power water pumping facilities may have both risk mitigation and cost saving benefits for public power utilities by enabling municipal water infrastructure to continue operating even during a grid outage.

Cooperative and public power utilities in Texas are showing how resilience can reduce costs with local, distribution-scale solar and energy storage. Because distribution-scale solar projects, which range in size from about 1 MW to 10 MW, are local and use local distribution systems, they help utilities reduce costs associated with distribution capacity and transmission. These economic benefits to local utilities, referred to as avoided costs, add up to savings of up to 50 percent of power procurement costs for local Texas utilities.[18] Texas cooperatives are also leading in the installation of utility-scale energy storage. In 2020, Pedernales Electric Cooperative, the largest cooperative utility in the States, installed a grid-scale battery storage system that benefits its members and the Texas grid, and will enable cooperatives to learn more about the benefits of incorporating energy storage into their infrastructure.

The same technologies that deliver energy resilience can also drive revenue from electricity sales. The Federal Energy Regulatory Commission's Order 2222 and follow-up Order 2222-A, issued in 2020 and 2021 respectively, opened up the wholesale electricity market to distributed energy resources like solar projects and energy storage systems, including the ability to sell electricity and to provide demand response services when utilities need customers to use less power during peak hours. Realizing this potential, however, will be complicated and will take time to resolve, due in part to the fact that the wholesale market was structured for big, billion-dollar power plants, not nimble microgrids and small-scale solar and energy storage. Permitting fees, engineering requirements, and decision-making timelines to receive approval to participate in the wholesale market can cost more than a solar and energy storage project. Patience, partnership, and policy—plus leveraging innovative-focused grant funding and incentives to cover these additional costs—are helpful approaches to overcoming barriers to change.

From Resilience to Restoration

Eric Clifton is the founder and CEO of Orison, a company that envisions "a revolution in energy storage that enables a world with abundant, clean, and affordable energy for all, where utilities and the people they serve work together for a brighter future." Clifton is also a man of faith whose commitment to build Orison was an unexpected answer to his prayer, which inspired the company's name. Orison is headquartered in Wyoming, where he lives in a rural community with his family. Clifton observes that a focus on control and centralization has been among the reasons it's been hard to move forward toward a future in which everyone, including "the least of these" in the most remote places,[19] has access to wirelessly connected communities with light, power, and commerce.

That same impediment is reflected in the physical infrastructure of our energy systems. "One of the core problems since the inception of the electric grid is time-of-use and fluctuating usage," explains Clifton. "Generation plants aren't designed to ramp up and down quickly or to deal with transient spikes. Centralized services can't address end-of-line issues where people are actually using the power." In contrast, Orison manufactures an energy storage system that is designed to work like an appliance—including being portable, able to move when you do— instead of treating energy storage as fixed infrastructure. By increasing flexibility, safety, comfort, and resiliency, Orison also reduces the end-of-line impacts on the grid to improve stability and efficiency. This approach doesn't just jibe with Orison's business model, it's fundamental to Clifton's sense of purpose. His vision is that scaling energy storage access in the United States will enable his company to expand equitable access to energy to places in Asia and Africa that have no access to electric light or power.

From Innovation to Demonstration

Clifton's company is just one example within a rapidly growing field that shows how we can move beyond our industrial past, which required centralization and control, and toward a technological future that enables energy systems to meet local, even individual, needs. It also presents an extraordinary opportunity to restore justice by using the advancement of energy resilience to make sure that we all have power, that the most vulnerable among us can keep the lights on, and that the money we save on infrastructure helps meet more of our communities' needs.

Eric Clifton, founder and CEO of Orison. (Photo by Tanya Clifton.)

As we have discussed, increasing energy resilience with local clean power, energy storage, and microgrids is at the innovative edge of the industry. While there are many technologies, approaches, models, and examples to inform your direction, there's no one right way to go. In addition, the best options for your community will be informed by your local policy and utility environment, and by the local impacts of climate change that may make the place you live more vulnerable to disruptions from floods, fires, and extreme weather. As you'll note from our examples, many were part of pilot programs that implemented first-of-a-kind resilience innovations. These programs can invigorate local energy markets—identifying and overcoming obstacles, sharing new knowledge across the system as a whole, and leading by example. Embrace the role of innovation, and begin building local energy resilience by identifying a demonstration project that showcases the value of resilience to your community. Even if it's a relatively small project—say, a solar and energy storage system on a church used as shelter during storms—your work will bring the possibilities to life and enable people to understand how resilience benefits your community. Much like the Community Resilience Hub program in Baltimore or the pioneering energy storage projects in rural Texas, you will be able to use what you learn through demonstration to make resilience a fundamental building block of your community's clean energy future.

CHAPTER 8

EVs

The Transformation of Transportation

The entire transportation sector—cars, trucks, buses, equipment, and all the fueling and other infrastructure that go along with them—is being electrified, not just in the United States but globally. This large-scale transformation of a major sector of our economy is big business that will have a correspondingly large-scale impact on our energy infrastructure. According to 2020 data from the US Energy Information Administration, the transportation sector accounts for 26 percent of all the energy consumed nationwide, nearly 80 percent of which is used by passenger vehicles and commercial trucks and almost all of which is fueled by gasoline.[1] When electric vehicles (EVs) eventually supersede cars and trucks powered by internal combustion, all that energy will shift from the fuel pump to the electric grid. As a result, US electric generation capacity will need to increase; our grid infrastructure will need to expand and modernize; and technologies that make our energy infrastructure more efficient and resilient, such as microgrids and energy storage, will need to deploy at pace to make the most of the benefits of electrification.

The transportation electrification journey is just beginning, and it is a massive undertaking. It will reshape and retool auto manufacturing; bring new manufacturing opportunities, including the production of batteries, to the United States; change how we maintain vehicles; increase the demand for renewable electricity; improve the electric grid with new battery capacity; and even transform our highway infrastructure. As we explore the electrification of transportation and what it can mean for your energy future, think about how your community connects with transportation—from its transit needs to local logistics and distribution centers to regional manufacturing. As we introduce examples of local electric transportation electrification, visualize the partnerships you can build to bring the value of transportation electrification home. You might work through your local nonprofit utility to electrify school bus fleets or launch an app-based electric vehicle transit system to help rural residents affordably get to work—there's no one right place to begin. The important thing is to get started. The opportunities for rural communities and small towns are enormous, but only if you engage.

The Story of "Two Rays"

Rey León is the mayor of Huron, California, a rural agricultural community where many farmworkers don't have sufficient household income to own a car. While the community has access to a regional transit system, it can take almost three hours by bus to travel fifty miles from Huron to Fresno, the nearest big city, for necessities like health care. As an alternative, residents can call a *raitero*—the Spanglish roughly translates to "a person who gives rides"—to take them where they need to go faster and with fewer stops than the bus. Common across California's Central Valley, the raiteros' informal, homegrown rideshare system meets community transportation needs, but at a cost of up to $100 per round trip.

In 2013, Mayor Léon, a longtime community leader as well as the founder and executive director of the Latino Equity Advocacy & Policy (LEAP) Institute, saw an opportunity to bring together the emerging market for electric vehicles and his community's transit needs by establishing Green Raiteros. His early fundraising efforts were frustrated by state institutions that didn't understand or value the importance of investing in equitable solutions for rural and low-income communities. Then, in 2016, the LEAP Institute connected with the Greenlining Institute and received the technical assistance it needed to win the project's first major funding award.[2] Green Raiteros debuted the following year, backed by an initial $519,400 of support. Equipped with two EVs and a team of volunteer drivers, the program made twenty-five to thirty trips per month, many to take residents to medical appointments. After an initial promotional period during which rides were free, Green Raiteros set its rates at $0.55 per mile—about half the cost of a trip with a conventional raitero. While launching and expanding the program have involved challenges—recruiting drivers, securing insurance, implementing a ride-scheduling dispatch system—Green Raiteros continues to grow, and it contributes an essential service to the local community.[3] In 2021, four Chevrolet Volts joined the Green Raiteros fleet on a special loan, thanks to a partnership between the LEAP Institute and General Motors to improve farmworkers' access to COVID-19 testing and vaccinations.[4]

The Ray, a living laboratory for sustainable transportation in rural Georgia, was conceived during a ceremony to name an eighteen-mile stretch of Interstate 85 through rural Troup County in memory of green industrialist Ray Anderson. During the event, Ray's youngest daughter, Harriet Anderson Langford, was struck with the realization, as she has described it: "I'd just put my daddy's good green name on a dirty

LAGRANGE

EXIT 14:
RIGHT-OF-WAY SOLAR

I-85

EXIT 6:
VEGETATIVE LAB

BIOSWALES

GEORGIA

ALABAMA

WEST
POINT

THE RAY
WELCOME CENTER

TIRE SAFETY CHECK STATION

SOLAR-PAVED HIGHWAY

SOLAR POWERED VEHICLE CHARGING

POLLINATOR MEADOW

THE RAY

SMART
STRIPING

RUBBERIZED
ASPHALT

Electric vehicle infrastructure and other sustainability focused pilot projects installed on The Ray. (Source: The Ray.)

highway, and I had to do something about it." As a result, Langford launched The Ray with a bold but simple mission: to set a new standard for roadways of "zero deaths, zero waste, zero carbon, and zero impact" by putting novel partnerships and emerging EV technologies to work in a different way.

Led by executive director Allie Kelly, The Ray works by demonstrating technology, sharing research, and informing future-focused policy—all through collaborative and sometimes unconventional partnerships. Among its first projects were a roadway paved with solar cells that produces enough electricity to power safety equipment, a 1 MW solar installation on five acres of interstate right-of-way, and an online mapping tool that shows all the underutilized roadside land that could instead be generating renewable energy—enough to power twelve million passenger EVs!

Working closely with state departments of transportation has been critical to The Ray's success. As EVs replace conventional gas- and diesel-fueled vehicles, state gas tax revenues will decline; the gas tax, which functions as a user fee, pays for roads and road maintenance. More EVs will mean less money for departments of transportation unless new sustainable funding models are developed that are aligned with new technologies. Electrifying roadways can be a source of revenue that replaces declining gas taxes, and The Ray's research gives state departments of transportation everywhere a reason to move forward confidently into an electric future.

The Ray already features PV-to-EV charging stations and envisions installing fast-charging lanes so EVs can charge while they drive. (Source: The Ray.)

The Ray has also partnered with leading corporations that are developing technologies to make roadways safer and more efficient, which has been a boon to local economic development efforts for the area's small towns. New and emerging projects are attracting positive attention and new businesses: a pilot autonomous EV freight corridor, for example, and a future commitment to testing EV fast-charging lanes that would enable people to charge their vehicles while they drive.

While the stories of Rey León and The Ray illuminate possibilities, the electrification of the transportation sector holds even more promise for rural communities and small towns. To help you envision how your community fits in, let's start by exploring the size of the market opportunity and a few of the policies enabling it, with a focus on how electrifying transportation can be done in America.

How Big Is the Opportunity?

Consider the number of passenger cars and trucks on US roads. Today, there are more than 270 million vehicles, of which about 1.6 million are currently electric. By Rocky Mountain Institute's analysis, if we use our shared carbon-reduction goals as a benchmark for determining the pace at which we must electrify the American fleet, we will need 70 million EVs on the road by 2030, with 100 percent of all new vehicles being EVs by 2035. In addition, we'll need about 300,000 fast chargers and 200,000 Level 2 chargers installed by 2030 to make sure everyone driving an EV has easy access to charging stations.[5]

That may sound like an extraordinarily big change in an extraordinarily short time, but remember how rapidly renewable energy is growing and keep in mind that the same large corporate buyers driving the renewables market are also committing to convert 100 percent of their fleets to EVs. Amazon.com, FedEx, UPS, and Walmart are among the biggest fleet operators in the United States with logistics operations that span the globe, and they have all adopted fleet electrification goals that include zero carbon emissions. Their leadership will not only move the EV market; it will also help commercialize new technologies, build scale, and bring prices down for the rest of us.

Policy at the federal, state, and local levels is enabling this transformation. From states like Massachusetts and California, which have set a goal of 100 percent EV sales by 2035, to federal investments in infrastructure, manufacturing, and R&D, governments are increasingly working in concert with local communities and leaders in the transportation sector to support electrification. It's not just about big policy; small changes matter, too, including changing arcane rules that preserve the status quo at the expense of the innovation. For example, the US DOT issued clarifying guidance in April 2021 about how funding from the agency can be spent that could increase available federal support for EV charging infrastructure by more than $40 billion.[6] Without this kind of helpful guidance, it wouldn't be clear if or how existing programs could support new technologies.

By 2050, electrification across all sectors of the US economy could increase electricity consumption by as much as 38 percent, with the largest increase driven by transportation. Electricity is purchased and distributed by utilities at the state level, and some states will experience more growth than others.[7] For many, the growth in electricity sales will be welcomed. For small town and rural cooperative utilities in particular, increased sales due to growing electricity demand from the transportation sector will be an opportunity to increase revenues available to reinvest in the local community. Electric vehicles give back in other ways as well, including through infrastructure that can increase the efficiency and resiliency of the local grid.

Make It in America

The opportunity, however, is not limited to vehicle conversions and the energy and infrastructure it will take to power hundreds of millions of EVs. It also includes the potential to make those vehicles in America by modernizing and expanding our manufacturing capabilities. This expansion of US auto manufacturing would include batteries, battery cells, drive trains, and assemblies, and it would extend to the transportation research and development enterprise and include the development of new maintenance and repair capabilities. To connect these global opportunities to local economies, America will need an industrial policy that makes US EV manufacturing a priority, rivaling the national government investments global competitors make in their own auto industries.

The US will have some catching up to do. In 2019, 325,000 EVs were made in America while 1.2 million were made in China and to date only seven of the forty-four automotive manufacturing plants in the United States have committed to full EV conversion by 2025.[8]

It's not just about the vehicles; it's also about the batteries. A massive EV battery plant built by South Korean SK Innovations is under construction in Jackson County, a primarily rural area in Northeast Georgia that's located about an hour from Atlanta, on I85. The first of two manufacturing facilities is expected to employ a thousand people and produce enough batteries to equip two hundred thousand EVs per year. The project will cost more than $2.6 billion to complete, but that's only part of the investment these facilities are bringing to Jackson County. Lanier Technical College is also investing in job skills training so that local members of the community are ready to go to work in this important emerging industry.[9]

Investing in people is essential, as my hometown learned when Kia built a large automotive manufacturing facility in neighboring West

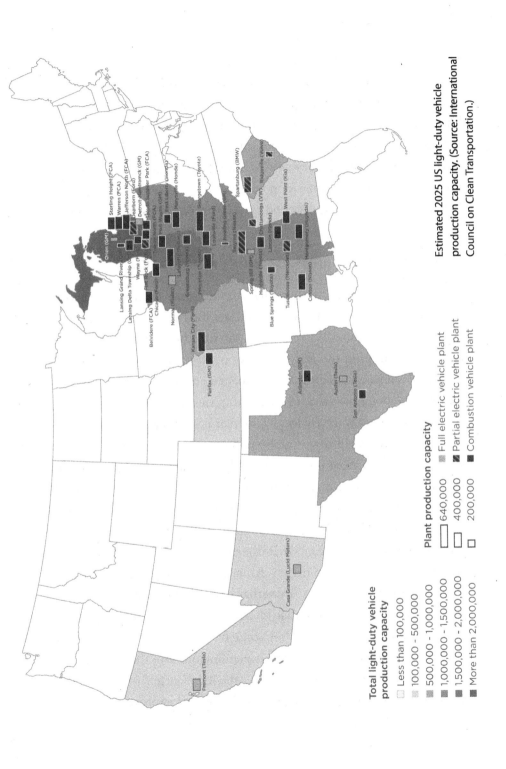

Total light-duty vehicle production capacity

☐ Less than 100,000
☐ 100,000 - 500,000
▨ 500,000 - 1,000,000
▨ 1,000,000 - 1,500,000
▨ 1,500,000 - 2,000,000
■ More than 2,000,000

Plant production capacity

☐ 640,000
☐ 400,000
☐ 200,000

▨ Full electric vehicle plant
▨ Partial electric vehicle plant
■ Combustion vehicle plant

Estimated 2025 US light-duty vehicle production capacity. (Source: International Council on Clean Transportation.)

Plant labels: Sterling Height (FCA), Warren (FCA), Jefferson North (FCA), Dearborn (Ford), Detroit-Hamtranck (GM), Flat Rock/Lafayette Park (FCA), Orion (GM), Lansing Grand River (GM), Lansing Delta Township (GM), Wayne (Ford), Flat Rock (Ford), Belvidere (FCA), Chicago (Ford), Normal (Rivian), Toledo (FCA), Wayne (GM), Marysville (Honda), Lafayette (Subaru), Greensburg (Honda), Princeton (Toyota), Louisville (Ford), East Liberty (Honda), Georgetown (Toyota), Spring Hill (GM), Smyrna (Nissan), Huntsville (Honda), Chattanooga (VW), Lincoln (Honda), Spartanburg (BMW), Ridgeville (Volvo), West Point (Kia), Montgomery (Hyundai), Blue Springs (Toyota), Tuscaloosa (Mercedes), Canton (Nissan), Kansas City (Ford), Fairfax (GM), Arlington (GM), Austin (Tesla), San Antonio (Tesla), Casa Grande (Lucid Motors), Fremont (Tesla)

Point in 2009. As former mayor Jeff Lukken explained to me, the community anticipated local unemployment would go down, but it initially went up as jobseekers from other regions moved to town and local members of the community struggled to meet Kia's hiring requirements for job skills. Once local communities, the state, and Kia came together to offer people the training they needed, more local people got jobs and the investment began to realize its potential for the community. You can't invest in technology and get good results without investing in people.

Local Benefits of Transportation Transformation

It can be intimidating to consider the transformation of the global transportation sector and its implications for the United States from the perspective of how to capture the potential benefits for your hometown. Keep in mind the early successes of Rey León and The Ray and how their efforts have grown in influence and impact, and focus on aligning the potential value of electrifying transportation with the specific opportunities you have locally. Remember as well that rural electric cooperative and small-town public power utilities are in a particularly strong position to grow alongside the electrification of transportation to the benefit of the communities they serve. For example, the growth of electricity demand not only increases electricity sales and revenue but also creates opportunities to expand renewable energy development and thereby tap two value streams that can support your community. To the extent that transportation-driven demand growth is the result of vehicle fleet conversions by companies that have adopted renewable energy goals, there is also the potential to build new public–private partnerships as The Ray did. Let's explore some of the potential opportunities, beginning with the individual and expanding out into the community.

Electrifying transportation has cost-saving benefits for individuals and families that can increase economic equity and improve affordability and quality of life. It's about $8,000 less expensive to own an EV than a conventional gas-powered vehicle over the vehicle's first two hundred thousand miles,[10] but many families struggle to afford the up-front costs of buying electric. In addition to innovative programs such as Green Raiteros that put the lower life cycle costs of EVs to good purpose to reduce transit and shared transportation costs for residents, utilities like the New Hampshire Electric Coop are making it easier for people with low and moderate incomes to access EV savings by offering rebates for new and used EV car purchases.[11] Including new and used EVs within the scope of rebate programs helps to expand affordability to more people with a broader range of price points. In addition, time-of-use rates can reward EV owners with lower electricity rates for charging EVs at specific times of day, which also benefits the local utility's demand management needs. The Sacramento Municipal Utility District, which had more than five thousand EV customers using their time-of-use rates as of 2019, has been an EV public power pioneer.[12]

Moving beyond the individual to corporate benefits, local economies can use transportation electrification to build new and expand existing corporate relationships. Rural and small-town communities are home to a diversity of industries, including many manufacturing and distribution hubs that depend on logistics operations to service supply chains and deliver goods. Offering transportation electrification as a service, including the infrastructure and renewable energy that are required by an ever-growing list of corporate leaders, is an opportunity to differentiate your local community and attract new investments from companies that are looking for locations that offer clean energy and transportation capabilities. These logistics centers, sometimes referred to as inland ports, can be a powerful competitive advantage. Just ask the Utah Inland Ports Authority (UIPA), a state corporation focused on generational planning for a safe, smart, and sustainable statewide logistics system. In May 2020, UIPA announced a clean energy cooperation agreement to plan for the electrification of freight, cargo, and logistics while meeting future needs with 100 percent renewable energy—which is a keystone of its plan for a thriving future.[13]

As we touched on in our discussion of resilience in the last chapter, building out charging infrastructure to support EVs can also provide new sources of revenue from grid efficiency, resiliency, and utility infrastructure cost savings. Also known as vehicle-to-grid charging, bidirectional charging not only enables EV batteries to store energy that can be used later but also allows EV owners to participate more easily in active demand response programs and other incentives that offer less expensive electricity to consumers in exchange for charging vehicles at times of the day when there is less overall demand on the grid. This approach, which uses technology to enable us to use and share electricity when it's best for everyone without compromising our individual needs, shares value across a connected community of EV drivers through local utilities that produce and provide electricity.

From a bigger picture perspective, while it's challenging to define its economic value, electrifying transportation also cleans up the air we all breathe so long as we continue to eliminate polluting fossil fuels and expand renewable and carbon-free sources of electricity at the same time. While the national and global impacts are compelling, EV conversions of gas and diesel fleets at the local level can deliver air quality benefits you can see and feel. For example, nearly 95 percent of America's school buses run on diesel fuel, the exhaust from which includes particulate matter and toxic pollutants that are linked to respiratory diseases, including asthma, and to cancer. Children breathe 50 percent more air per pound of body weight than adults, so the impact of breathing in school bus diesel exhaust is particularly harmful to kids. Converting America's 480,000 school buses to electric vehicles could have a profoundly beneficial impact on children and their health. Electric school buses also support good clean jobs, thanks to three American EV school bus companies: Blue Bird in Fort Valley, Georgia; Proterra, with offices in California and South Carolina; and Thomas Built Buses, which part-ners with Proterra to build EV buses in High Point, North Carolina.

The Value of Partnership

Building partnerships is at the core of many successful strategies for bringing the benefits of the electrification of the transportation sector to your community. Local utilities working with school districts, local economic development teams working with corporate logistics leaders, and nonprofits working with municipalities to create transit solutions are just a few examples of collaborations that can result when EV infra-structure is tailored to local communities' priorities. The scale of these opportunities is too great to go it alone, you can typically realize a vision faster together, and building partnerships is essential to building an EV strategy that will work for your community. There are many good, local examples to follow.

Clean Energy Works, the same nonprofit advisors that helped to scale the Pay As You Save (PAYS) model for energy efficiency, are also pioneering scalable approaches to electrifying transportation using the same model. Margarita Parra leads the Clean Energy Works transpor-tation portfolio in the United States as well as in Latin America and prioritizes energy equity across all her work. Noting that the economics of transportation electrification are different in urban centers, where the opportunities and economics are driven by density and the sheer size of the market, Margarita focuses on the opportunities unique to rural communities. As she explains, "Vehicle electrification is an opportunity to bring business to rural utilities. Examples include rural transit, school buses, and fleets. There are also services that can be offered through electrification, including microgrids and vehicle to grid. We don't yet know what value those streams will offer, so we're focused on studying and understanding the potential value and how it can be shred equitably through cooperative utilities."

Margarita relates the example of a suburban county school district that successfully used an energy performance contract to transform its 1,400-vehicle school bus fleet. Instead of paying for new school buses up front, Montgomery County Public Schools in Maryland entered into a contract with Highland Electric Fleets to subscribe to electric school buses as a service through an operational lease for the same cost that the school district pays to purchase and operate its fleet of diesel buses. By 2023, the school district anticipates it will not buy any more diesel vehicles and will be well on its way to converting its entire fleet to EVs.[14] Providing school districts with a turnkey EV school bus fleet without increasing school budgets is a financing solution that could help many more districts make the same change.

Clean Energy Works is also researching how a similar approach using PAYS could work for rural cooperative and small-town public power utilities to provide transportation electrification as a service. Using this approach, local nonprofit utilities would not only benefit from additional electricity sales but also grow by offering new lines of business that benefit their members. Always at the forefront of energy innovation, Curtis Wynn sees additional opportunities to capture the value of using school bus and other EV batteries for energy storage. In this scenario, idle school buses could store energy during school vacations which could be used on demand by the local utility when the demand for electricity peaked, such as on hot summer days.

Local utilities are also partnering with state and local government to expand fast-charging infrastructure, removing a major barrier to transportation electrification. In February 2021, the Tennessee Valley Authority (TVA)—the nation's largest public power utility—announced a partnership with the Tennessee Department of Environment and Conservation to develop a statewide network of fast-charging stations with locations at least every fifty miles along interstates and major highways. Funded in part by Tennessee's Volkswagen Diesel Emissions Settlement and Environmental Mitigation allocation—funding that every state received—the program is expected to cost about $20 million.[15] The Tennessee Valley is one of America's fastest-growing regions for automotive production, and the TVA's president and CEO, Jeff Lyash, has made a point of committing TVA as a partner in regional EV leadership. "TVA has reduced carbon emission by nearly 60 percent since 2005 and we have concrete plans to reach 70 percent by 2030," commented Lyash. "Actively supporting the electrification of transportation multiplies our own carbon reduction efforts and moves the entire region toward greater sustainability and economic opportunity in the future."[16]

Finally, local entrepreneurs are bringing home the benefits of transportation electrification through partnerships. Francis Energy once was Francis Oil & Gas, a family-owned Oklahoma company founded in 1934 by Sam Miller. Miller's great-grandson, David Jankowsky, went back home to Tulsa ninety years later to "bring clean energy to the heartland" under the family name. Founded in 2015, this generation's Francis Energy develops fast-charging infrastructure for EVs instead of developing oil and gas fields. Focused on serving rural communities, the company describes its mission as "enabling the acceleration of the EV market." Francis Energy has installed more than five hundred fast-charging stations across Oklahoma, including EV infrastructure serving the Cherokee Nation to power its first EV buses.[17]

Federal investments in modernizing US infrastructure also expand funding opportunities and show the power of partnership. Many federal programs—including funding available through the Departments of Transportation and Energy, the Environmental Protection Agency, and the lesser-known Economic Development Administration within the Department of Commerce—require that municipalities or other units of state or local government be either the applicant or a partner in the project team seeking funding. Electrifying school bus and transit bus fleets, developing new EV-based rideshare programs for rural communities and small towns, and investing in people so that those who want to work in an electrified transportation sector have the skills they need are all the types of local programs that could be jump-started with a values-aligned partnership and federal infrastructure funding.

"The Power of One"

The electrification of the transportation sector is a large-scale transformation that will also drive changes in the electric power sector in the United States and globally, accelerating demand for renewable energy. This global transformation holds local opportunities that you and I can help our communities reach. As Ray Anderson, namesake of The Ray, always explained, one person can change their community and the world by changing one mind at the time.

Whether the priority for your community is developing bidirectional EV-charging infrastructure, electrifying the school bus fleet, or training people to earn a living manufacturing and maintaining EVs, there are a multitude of good places to start. Your commitment can get the ball rolling, but the opportunities are too big to do it alone. Each of the stories of leadership and innovation we've shared began with one person's vision but succeeed because of a team's efforts.

CHAPTER 9

Rural Broadband, Smart Grids, and Rural Power

We've now covered four primary approaches to building a local clean energy future: energy efficiency, solar at every scale, resilience, and electronic vehicles and electrification. Now it's time to tie it all together with broadband internet. Broadband is intertwined with rural power both technically and historically. It is a necessary component of a smart, modern grid, and yet access to broadband service is extremely limited in rural communities for many of the same reasons that access to electricity was lacking a century ago. As broadband moves from being a luxury for those who can afford it to a public utility that's necessary for quality of life and economic growth, rural cooperatives and public power utilities—the same local institutions that brought power to farms in the 1930s—are in ideal positions to expand broadband access by extending the fiber networks they're already building to modernize the grid.

In this chapter, we'll introduce broadband's role in clean energy, including why electric utilities are already deploying broadband for their own use. We'll then examine the parallels between the history of rural power and the contemporary lack of broadband access for rural communities, including why the digital divide has persisted despite an influx of infrastructure funding. Finally, we'll discuss what it would take for local nonprofit utilities to be able to take the lead on broadband, and why everyone who envisions a clean energy future for their community needs to become a rural broadband advocate too.

Why Smart Grids Need Broadband

Broadband is a fundamental element of a smart distribution grid, which enables two-way communications and data exchange across a network of electricity generation, distribution, storage resources, and the homes, buildings, and appliances that use energy. Broadband-connected smart grids allow all their elements to work together responsively. The resulting heightened need for two-way communications between utilities, smart meters, smart appliances, microgrids and batteries, bidirectional EV chargers, and other technologies and equipment has vastly increased the volume of data and communications necessary to maintain the reliability of our energy systems while also improving efficiency and performance. This digital transformation of the electric utility sector follows similar transformations of other industries, like manufacturing, during which new technologies (robotics, for example) were introduced that increased productivity but required more computing capacity and bigger, faster communication networks to support operations.

Our old energy infrastructure didn't contemplate homes, businesses, and neighborhoods that would be able to generate and store their own power or smart appliances that could be responsive to the demand management needs of utilities. Communications needs were limited to connections across the utility's own equipment and infrastructure, and copper wire—the same technology used for dial-up internet connections—was sufficient to meet the need. Technology, however, has advanced.

Utilities today need vastly expanded communication capabilities to build and manage smart grids. Some of these requirements, many of which are directly related to building clean energy futures, include integrating distributed renewable power generation and storage, increasing resilience against cyberattacks and natural disasters, increasing operational efficiency, enabling new energy services, creating opportunities for customer participation in demand management, and enabling a "self-healing" grid that can automatically mitigate power outages. All these complex communications also generate an extraordinary volume of data, including energy usage data from smart meters and appliances, which has expanded utilities' needs for maintaining data privacy and increasing cybersecurity.

For residential customers, the results of smart grid technologies, such as appliances that use power only during times of the day when it's the cleanest and cheapest, include more reliable and affordable electricity. While much of the work to achieve these outcomes happens on the utility's side of the meter, behind the scenes, some of the experiences you might have as a customer with a cleaner, smarter grid could include getting credits or rebates on your utility bill for charging your EV at specific times of the day, participating in demand management programs to help manage peak usage, or signing up for a community solar subscription.

Smart grid services require high-speed network connections with sufficient bandwidth to transmit large volumes of data extremely rapidly and with great reliability, which means that smart grids need broadband. While utilities once leased network access from telecommunication companies, they're now increasingly running their own optical fiber, which is among the fastest and most reliable types of broadband technologies, able to transmit data over very long distances up to 70 percent as fast as the speed of light. This approach not only meets the infrastructure requirements of a modern grid but also responds to important differences between how telecommunications companies and internet service providers (ISPs) operate. While utilities are required to serve everyone with electricity and reliability, telecommunication companies and ISPs serve areas with the highest density of customers, neglect areas with low densities of customers, and aren't held accountable for reliability. Moreover, utilities have expressed concerns that they would not be able to depend on priority access to private broadband networks in an emergency.[1] Consider this from the perspective of your own experience: How often does the power go out compared to how often your internet or wireless service wobbles or goes down? Does your cell phone work during an emergency, or have you found that "all circuits are busy"?

These differences boil down to the distinction between a public utility, like electricity, and the associated regulatory requirements that cooperative, public power, and investor-owned electric utilities share to serve everyone reliably, versus a service that has heretofore been a luxury, such as broadband, which is available only to those who can afford it in places that are profitable for privately owned broadband providers to serve. Broadband access, however, is quickly becoming viewed as a public utility that is necessary for people to participate in education, commerce, and health care, a reliance that has intensified during the COVID-19 pandemic.

This convergence of broadband's evolution into a public utility, and electric utilities constructing broadband networks to modernize their infrastructure and build a smart grid, creates an opportunity for existing electric utilities—especially cooperative and public power utilities—to provide broadband access. These same local nonprofit utilities that were created to marry electrification with economic development can do the same with broadband as an integral part of building a clean energy future. And yet, there are familiar obstacles.

Rural Broadband and the Lack Thereof

The Federal Communications Commission (FCC) defines broadband as 25 megabits per second (Mbps), which is a sufficiently high-speed internet connection for one person to watch a high-resolution 4K show on Netflix, or for two to three people in a household to be on the internet at the same time. In many dense urban and suburban communities, ISPs including Xfinity, Verizon Fios, and AT&T offer connection speeds that are hundreds or even a thousand times faster. In rural communities, however, 19 percent of households and many businesses have no access to broadband and suffer with DSL internet connections that are less than a third as fast.[2]

Data from the National Telecommunications and Information Administration indicates that forty-two million Americans, including both urban and rural households, lack broadband access. As a May 2020 study

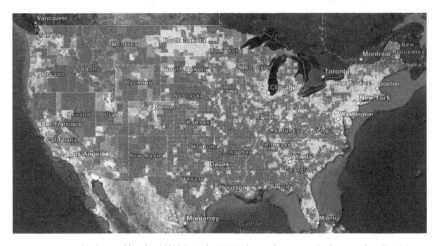

A mapping tool released by the NTIA in July 2021 shows how much deeper the digital divide is in America than the FCC's maps would indicate. Darker areas indicate less or no broadband access. (Captured from https://broadbandusa.maps.arcgis.com/apps/webapp viewer/index.html?id=ba2dcd585f5e43cba41b7c1ebf2a43d0.)

on Baltimore from the Abell Foundation noted, broadband access is a function of both infrastructure and economics.[3] Household income represents the greatest digital divide, which is not surprising given that broadband access, particularly at higher speeds, remains a premium service. While 43 percent of adults earning less than $30,000 per year lack access to broadband, only 7 percent of adults earning $100,000 per year or more are missing out on high-speed internet.[4]

In the case of rural America, however, which has proportionally less access than urban communities, broadband is a "last mile" infrastructure challenge. Very much like the early days of electrification, when private utilities refused to run power lines to farms because it wouldn't be profitable, private telecommunication companies and ISPs won't run fiber to many rural communities. While actual costs vary from place to place based on local conditions, expanding broadband networks into rural communities is estimated to cost from $25,000 to $50,000 per mile.

The costs of not expanding broadband infrastructure are high too. As Mayor Don Hardy of Kinston, North Carolina, has candidly observed, "Companies aren't going to where the Internet sucks."[5] Rural and small-town communities with no broadband access not only struggle to attract and sustain new employers but also suffer from a lack of educational access. During 2020 when the COVID-19 pandemic sent kids everywhere home for remote school, an estimated fifteen million children had no access to high-speed internet,[6] which meant no Zoom classrooms and limited access to online textbooks. I could see the impact in rural communities near where I live on school nights, when library parking lots would fill up with parents and their children accessing public hot spots to do homework for school.

While the federal government has been investing more and more money into expanding broadband since the Telecommunications Act of 1996 included it in the definition of "universal service,"[7] the broadband gap still lingers. Billions of dollars of public funding and incentives have been granted primarily to private telecommunications companies to improve access, but the "last mile" infrastructure that's necessary to increase access for rural communities is not yet in place.

Parallels with the Story of Rural Power

Rural communities are being left without high-speed internet access for many of the same reasons they were left without electricity until the 1930s, when the combination of action by the federal government and organizing by local communities to form cooperatives and public power utilities changed the electrification landscape. Private companies dominate the broadband market and assert that it's too expensive to provide service to less densely populated areas. Because broadband has only recently begun being viewed as a necessity, there has little regulatory pressure to close the gap, though billions of dollars are being spent in an effort to incentivize companies to provide service. The key difference, however, is that the rural electric cooperatives and public power utilities that were created to turn the lights on for rural America in the 1930s are already well established and providing other, highly related energy services to the same communities. While big companies like AT&T and Verizon are considered the incumbent ISPs, these local energy democracies are the incumbent public utilities. Moreover, they are already using their existing infrastructure, including utility poles and rights-of-way, to run fiber.

Also like the story of rural power, the big private telecommunications companies are using their influence to prevent competition and maintain tight control of markets for themselves. In fact, in some states, telecommunications lobbyists have successfully changed laws to prevent public power utilities from providing residential broadband access while also requiring that publicly funded fiber networks be made available to private companies at very low costs. Nationwide, at the time of this writing, eighteen states have laws in place that restrict community-owned broadband networks, including states as diverse as Florida, Minnesota, and Nevada.[8] Restrictions range from outright bans to other, more nuanced, regulations that, for example, prohibit rural cooperatives from using USDA's low-interest loan programs to deliver broadband service.[9]

This situation has been aggravated by flaws in how the FCC, which has tremendous influence over the broadband market nationally, implements rural broadband regulations. For example, the FCC's broadband access maps have understated the broadband access gap by at least half, in part because the FCC depends on self-reported data from telecommunications companies, which has historically been unreliable. While the FCC may state that a rural census tract has broadband access, the reality may be that only one or two homes in the area are served. The June 2021 release of broadband access maps by a different federal agency, the Federal Telecommunications and Information Administration (FTIA), underscored the issue. FTIA's maps used broader sources of data that include information on network performance rather than self-reported information from incumbent internet providers, and showed that 42 million Americans were without broadband, not 18.3 million as the FCC had reported.

The FCC's flawed broadband maps, rooted in their linkages to self-reported data, are of critical importance because the FCC's maps determine how significant amounts of federal funding to improve broadband access are distributed. If a rural community, defined by census tracts, has broadband access according to the FCC's erroneous maps, then it's not eligible for key federal programs. The extent of these issues, however, is even deeper.

Funding for building rural broadband is allocated by the FCC through a reverse auction, which means that companies bid for federal support to provide service to specific census tracts based on their estimated cost of service. The lowest-cost bid for service in the reverse auction wins

FCC funding to provide service to that census tract. The vast majority of the organizations that have won FCC funding are large, private telecommunications companies, including new and largely untested technology companies that are owned by billionaires, including Jeff Bezos and Elon Musk. For example, Elon Musk's company, which intends to deliver high-speed internet access with low-orbiting satellites, won more than $900 million in federal incentive funding in a recent FCC reverse broadband auction. These big companies underbid the cost of providing service, sometimes by more than 90 percent, to block other providers, including local nonprofit utilities, from delivering broadband to their member-owners and residents. Then they squat on the territory without actually expanding access for local residents. In short, billions of dollars in public funding are being given to big private companies that are using the funding process to block local utilities from providing broadband for the rural communities they already serve.[10]

Roanoke Electric Cooperative's experiences illustrate the issue. Through Roanoke Connect, its cooperative broadband service, residents in the rural, predominantly low-income service area are able to purchase Essential Home Connect broadband service, which offers internet connection speeds up to four times what the FCC defines as broadband, for $59.99 per month. However, Roanoke Connect is not able to expand service to the majority of its service territory, because the FCC maps either inaccurately show that residents already have service, or the FCC has auctioned off large portions of Roanoke Connect's territory to big private telecommunications companies that still do not provide service.

The FCC's problems also impact the states. North Carolina, like many states, offers grants and other funding to expand broadband access. Unfortunately, its Great Grants reference the FCC's maps and funding processes and prohibit any rural areas from participating if a telecommunications company has already received FCC funding to expand service there. As a result, the same rural communities that are misrepresented and unserved are also being excluded from state programs.

Broadband Connections and Clean Power: Local Solutions That Work

Despite the challenges, local nonprofit utilities are leveraging the broadband infrastructure they're building to enable smart grids and clean energy to provide residential internet services. It's essential to understand the potential to meet both needs at once and take advantage of the role that broadband plays in smart grids and clean energy. By building a clean energy future that's connected by broadband, you can do even more to help your hometown thrive.

In Virginia, two of the state's largest utilities, Dominion Energy and Appalachian Power, are partnering with local utilities to deliver residential broadband where it's lacking. By enabling local utilities to share access to the broadband networks these large utilities are deploying to modernize the grid, these utility partnerships are reducing the costs of rural broadband deployment, thereby making local service more accessible. Utilities have to serve everyone, and their infrastructure reaches everywhere, including remote rural areas,[11] so this pilot program from Virginia could work in many other states. Of note, this program was possible only because Virginia's Grid Transformation and Security Act of 2018 established a rural broadband pilot that enabled investor-owned utilities, including Dominion Energy, to work with cooperatives and other rural distribution utilities to expand rural broadband access. Without this legislation, private telecommunications companies in Virginia would have maintained their chokehold on the rural market. New legislation is where many other states will need to start as well. As Republican public service commissioner Jeremy Oden of Alabama has stated, "We must face it as we did electrification in the early 1900s. Rural internet and internet altogether needs to be a necessity now and not a luxury."[12]

Georgia offers another example. In 2019, Georgia passed a new state law making it clear that rural cooperatives can offer broadband services, which was a critically important market signal, especially contrasted against the exclusionary laws in other states. Georgia also developed its own broadband maps, which highlight the gaps in service access compared with what the FCC's maps have to say.[13] As a result, more rural cooperative utilities are leveraging their broadband networks to provide much-needed access across the state. For example, in March 2020, Diverse Power, an electric cooperative utility in Georgia, became the first in the state to receive approval from the Public Service Commission to offer residential broadband services through its affiliate, Kudzu Networks. As Diverse Power's CEO, Wayne Livingston, explained, "It is my belief that internet access is no longer a 'want' but rather a 'need.' Our board of directors, who live and work in rural communities, thought it important enough to step up and play a role in providing what is becoming an essential service for our consumer-members and communities to remain competitive and vital in the twenty-first century."[14]

Other examples recall some of the early histories of public power. The Mason County Public Utility District in Washington State will run broadband infrastructure to the homes of residential customers who sign up for a fee of $3,500 per home, which the utility district will finance over twelve years. The City of Ammon in Idaho enables residents to band together in neighborhoods to sign up for broadband fiber. By working together, residents can reduce the cost of broadband deployment for everyone.[15]

Your Role in a Connected Clean Energy Future

Clean energy needs broadband, and small town and rural communities need broadband too. Your local cooperative or public power utility can take the lead to make it happen by bringing together the broadband infrastructure that they are building to enable the transformation of our energy systems with the expansion of rural broadband access. This commonsense approach not only makes use of the infrastructure and institutions we already have in place in rural communities but also reduces the overall costs of rural broadband by deploying clean energy infrastructure to deliver multiple services.

This convergence won't happen on its own, however. It needs you and your community to become advocates for broadband. While private telecommunications companies and nonprofit utilities both have important roles in delivering universal high-speed service to all the diverse geographic regions of our country, like in the 1930s, corporations and their lobbyists are fighting to limit local competition. States like Georgia have already made it clear that local nonprofit utilities belong in the broadband delivery business, and pilot programs in places like Virginia are showing how utilities can work together in partnerships to deliver rural internet even more efficiently. While the FCC's flawed maps and rural broadband funding processes remain an obstacle, local and state leadership can help break those roadblocks as well.

Taking into consideration our lessons from the past, and incorporating your vision of all the ways in which clean energy and the broadband infrastructure that support it can help your hometown enjoy a thriving future, expand your engagement starting at the local level. While advocating for your community's needs may feel less concrete than building a local solar project or installing EV charging infrastructure, it is just as important. Organizing customers at the local level to stand up locally controlled utilities to meet local needs is exactly how the country was electrified.

CHAPTER 10

Scale without Losing Local Control

Now that you understand how you can use localized clean power to help revitalize your hometown with energy efficiency, solar power, resilience, electric vehicles, and broadband—all of which can be implemented through your local nonprofit energy democracy—it's time to talk transformation. Knitting together the good work of thousands of small towns and rural communities could be transformative in ways that extend well beyond the immediate benefits of clean energy. From improving the quality of rural housing with energy efficiency to retaining the wealth produced by energy resources by building solar to using the modernization of the electricity grid and delivering broadband to every home, propagating local clean energy futures across the country could sustainably reverse rural America's fifty-year decline.

As we consider how to scale our rural renaissance so that every hometown can thrive, we've got to keep in mind that localization is at the heart of how and why we can revitalize our communities using clean power. Scale shouldn't come at the cost of diversity or local control. The big, centralized physical, financial, and power infrastructure of our nineteenth-century energy system is at the root of the energy poverty, environmental degradation, and climate change we experience today. Our goal is not to re-create the same structures with new technologies. We want to go big in terms of positive impact but keep our priorities local.

While the independent democratic governance and locally oriented nonprofit business models of our nearly three thousand rural cooperatives and public power utilities make them ideally suited to build clean energy futures, their sheer number creates an obstacle to achieving that vision. Sharing the information, program models, tools, and technologies each community will need across such a large and diverse network takes more time and effort than driving a top-down, one-size-fits-all solution through a corporate chain of command. The enduring improvements to local quality of life are worth it, but how do we solve the inherent distribution problem so that we can get there?

To enable clean energy futures by sharing capacity without diluting local control, we'll focus on the "back office" infrastructure of clean energy, including finance, training, information systems, analysis and assessment, policy, and building community. By differentiating the building blocks of local clean energy futures that can benefit from a more national approach from the prioritization, design, and decision-making infrastructure that must remain local, we can equip local communities with the tools they need and not violate our values by telling them what to do. This allows us to build momentum that can transform energy futures everywhere.

Finance

Let's start with finance, because access to capital is a basic necessity for any type of clean energy project. The finance and banking industry denotes scale in billions of dollars, which I did not understand until

2011, when I had a conversation with a banker about financing energy efficiency. I was serving in the Obama White House, and we were just about to announce a new initiative to invest $2 billion to retrofit federal buildings to make them more energy efficient. The initiative would be paid for with energy savings, using a financing mechanism called an energy performance contract, which is similar to the Pay As You Save approach for home energy efficiency that we've discussed, but designed for commercial buildings. I asked the banker how a $2 billion commitment might make the same financing mechanism more available to state governments and the private sector, and perhaps help reduce financing costs, by creating a large-enough market to reduce the perception of risk and to provide more efficiencies from scale. He replied that $2 billion was too small to make much of a difference. You can imagine my surprise, but I took his response as an important lesson.

When it comes to making financing more accessible and less expensive to more communities, scale matters. The more familiar banks and other financial institutions become with technologies and solutions like on-bill energy efficiency programs and community resilience centers, the less risk they perceive in lending. Borrowing money at lower interest rates thus becomes more available and accessible. Financial products that help build clean energy futures, including loans and bonds, can scale very rapidly. Green bonds, for example, were first introduced in 2007 and now represent a $1 trillion global market.[1]

There is already an established group of financing options available to support the projects and programs you need to build a clean energy future, which includes the US Department of Agriculture (USDA) and National Rural Utilities Cooperative Financing Corporation loans we discussed in our chapter on energy efficiency. These programs already have many tens of billions of dollars of financing capacity, so achieving scale will require using the existing program capacity to finance clean energy so that it becomes commonplace, and therefore more efficient and accessible.

Keep in mind, it usually costs more and takes more time the first time you do something. So you may consider seeking philanthropic or other kinds of grant or sponsorship support to help cover the additional cost of innovation for your first clean energy project. From that point forward, however, programs like USDA's Renewable Energy Savings Program can help you get started, and higher-capacity financing solutions (municipal green bonds for public power utilities, for example) can help you, and others, scale the work.

Workforce and Training

Rural America needs a lot of clean energy practitioners to build the projects, run the programs, and manage the new businesses that will be created by the localization of our energy systems as we build more clean power. For a small community, training—or retraining—a workforce can seem like both a wonderful economic opportunity to support good new jobs and an overwhelming task.

Fortunately, you're not on your own. There are already many nationally accredited workshops, educational curricula, and certification programs to help train the small army of energy assessors, weatherization contractors, heating and cooling technicians, solar installers, electrical contractors, and other energy leaders that your community will need. Federal and state governments, academic institutions, and trade associations that serve the energy sector have been investing in clean energy education and professional certification processes for decades, so the workshops and certification programs your community will need are available. Moreover, many of them offer helpful opportunities for local tradespeople to distinguish themselves in their field through specialized credentials and recognition.

For example, the Residential Energy Services Network offers a training and certification program for Home Energy Rating System Raters (HERS Raters), which distinguishes individuals who have the building science expertise to test, diagnose, and identify all the energy efficiency improvements in a home that will reduce energy usage and utility bills. Becoming a certified HERS Rater can be a great way for a local contractor to build their business or for a residential real estate appraiser to expand the services they can provide to their customers. Similarly, national training and certification programs are available for solar installers, electric vehicle technicians, and many other careers people can pursue in a clean energy future.

Like finance, the tools to train a clean energy workforce for good local jobs that will support thriving communities already exist. Building scale can be achieved by deploying these established tools at the local level through community colleges, business-led professional development programs, local unions, and other institutions that invest in people to help them build knowledge and wealth through their work. To get started, reach out to the national trade association that represents your field of interest, like the North American Board of Certified Energy Practitioners for solar installation. They can guide you to the training resources you need. You can also reach out to the local leaders profiled in this chapter and throughout this book. They are all early innovators who had to figure things out for themselves, and the organizations they lead welcome the opportunity to help others by sharing what they've learned.

Information Systems

Information systems are essential to building the kind of scale that can help restore the vitality of rural communities everywhere. The information systems that support clean energy futures include everything from the software that manages your local microgrid to the customer-facing applications that let you control your home's thermostat while you're away. This constellation of energy information systems manages large volumes of confidential data such as your utility account number and home energy usage profile, so it's important that the energy information systems you use are safe and cybersecure.

Among the impacts of the digitization of the energy industry is that there's usually "an app for that," regardless of the application you have in mind, so there are already a host of software and other information tools to enable local clean energy projects and programs. That said, however, there is an extraordinary opportunity for additional customer-centric innovation in the development of energy information systems designed around how people use information instead of focusing narrowly on technical requirements.

One example of an energy information system that is ripe for customer-focused, potentially disruptive innovation is your electricity bill. If national trends hold true where you live, it's very difficult to understand your utility bill. You see how much energy you used only at the end of the month, when it's too late to do anything about it, and it's hard to discern what all the rates and fees on the billing statement actually mean. Creating new information systems, such as new utility customer account management applications that are designed to engage customers in their energy futures, holds the potential to help local clean energy solutions scale faster by providing better service.

Groundswell took this approach when we built our SharePower platform for managing community solar subscriptions. Our priority was and remains serving low- and moderate-income customers, so we involved these customers directly in the product design process. We made it a priority to understand how software design decisions impact people who may struggle to make ends meet as opposed to those in affluent households. For example, Groundswell's lower-income community solar customers were more likely to be over fifty-five years old and less likely to use email. By contrast, more affluent, market-rate customers preferred email communications and almost exclusively used their mobile devices for account management. As a result, customer communications for lower-income customers were designed to prioritize phone calls supplemented with conventional mail, and to incorporate text messaging if customers opted in and shared their cell phone numbers.

Putting people first and taking the time to understand the specific needs of different customer groups also drove design decisions such as how Groundswell allocates solar savings. For example, most community solar programs give everyone who subscribes the same savings, regardless of their income, which might result in 10 to 15 percent electricity-bill savings per month across the board. That means that if your monthly electricity bill were $150, you'd receive $15 to $23 in solar savings by participating. By contrast, Groundswell's SharePower program prioritizes savings for low- and moderate-income households, each of which receives a community solar subscription that covers about half its monthly electricity usage at no cost, thereby eliminating administrative barriers like consumer credit while saving each household about $500 per year. Market-rate community solar customers make this possible by subscribing at full price so that all available solar savings can be shared with lower-income participants. Using local solar power and helping your neighbors is more valuable to many than $15 per month, which shows in strong customer engagement and positive customer survey results.

While you will find many established energy information systems and software options to support your work, keep a lookout for the right opportunity to put the people you serve at the center of the solution and build your own. The EV rideshare service that Rey León founded in Huron, California, initially depended on the telephone and outdated scheduling software, but there's no reason that a similar approach developed specifically to meet the needs of rural communities couldn't replicate the success of Uber, Via, or Lyft. Customer-oriented innovation can be a competitive advantage, and while you might start at the local level, there are thousands of communities much like yours that could benefit from the same ideas.

Analysis and Assessment

Energy is a technical field, and standardized approaches for analysis and assessing program performance are a must. Adopting established approaches not only helps save time because you don't have to reinvent the wheel but also enables you to keep score and measure progress in ways that make it visible when another community is doing something a little better, which can help you improve too. This doesn't mean taking a one-size-fits all approach, but rather applying national standards that you can adapt to your own local needs.

You can help build scale to support established analytical and assessment frameworks by sharing data and participating in research studies. Both these ways of engaging create knowledge and support learning networks that improve access to information that other local communities can use in ways that are appropriate to their needs.

Solar and energy efficiency both provide great examples. In the solar industry, the Solar Energy Industries Association supports an annual study of industry diversity that depends on the responses of member organizations as a basis for research. The 2019 report identified a gender pay gap in the solar field that was even worse than the country as whole: women in solar earn just seventy-four cents for every dollar a man is paid for the same job.[2] Participating in and following this kind of industry-wide research gives you the knowledge you need to make sure your values are represented in your clean energy future. Being aware of a gender pay gap lets you know you need to ask the solar companies you work with whether they treat women fairly with equal pay, and make a change if they don't.

Scaling standard technical assessment tools, as well as sharing data, also helps build local energy futures while supporting scale across the field. For example, every energy efficiency program in America can use the same energy assessment methodologies and tools to measure home performance. How those tools are applied, however, varies from place to place. Hotter climates with more humidity may need different kinds of energy improvements than homes in cold or dry climates, and local variations in energy prices will impact which improvements will save more money than they cost. Sharing data and results from your program to support industry-wide research may seem inconsequential, but it can help other communities like yours have the confidence to move forward with energy efficiency. Just as Mark Cayce shared his experiences in Ouachita, Arkansas, with leadership in LaGrange, Georgia, collegial advice paired with data is a very helpful combination that supports program replication and scale.

Policy

As we've learned, decisions about energy policy are made at every level of government—from Washington, DC, to your hometown. Sometimes, sharing good ideas from one state or community to another can also help the market as a whole to reach greater scale. This is particularly true where policy obstacles exist because previous generations of policy makers didn't anticipate the technologies we use today.

Working at the local level to scale vehicle electrification everywhere is a great example. Many electric vehicle technologies are still emerging, and will require our existing infrastructure to perform in new ways. As we've learned, The Ray in Georgia demonstrated how highway rights-of-way can be used to develop solar energy projects that may one day enable PV-to-EV fast charging while you're driving. In order to build a 1 MW demonstration project, however, The Ray had to go through an intensive process to get permits and approvals, because no one had ever tried to build solar alongside a highway before. Thanks to their success, other states can now copy their work, including the state policies that were updated to recognize new technologies.

Building Community

The finance, training, information system, and other kinds of tools and infrastructure that are necessary for achieving transformative scale already exist. Moreover, how these solutions are applied at the local level, particularly when combined with customer-centric innovations that put people first, will enable you to implement clean energy programs and projects in ways that respect local priorities and preserve local control. Scale can be achieved at the level of shared services while retaining flexibility and autonomy in how those services are used to integrate with local energy systems to meet local needs.

As we build scale together across the country, keep in mind that relationships are infrastructure too. Whether you need advice from a friend on a thorny technical problem or a referral to find the right financial partner, or if your state had a policy breakthrough that you want to share with others who would benefit, building and contributing to a community of practice is part of how we can get big things done.

In part because energy connects everything, numerous professional associations, nonprofit organizations, and interest groups exist to bring like-minded people with common interests and goals together around state and local priorities, policy interests, technical topics, research, and other activities. Membership organizations like the National Association of Regulatory Utility Commissioners, regional nonprofits like the Southeastern Energy Efficiency Alliance, and state-level groups like Kentuckians for the Commonwealth are all examples of associations you can plug in to. Most important, get involved in your local nonprofit utility and with colleagues participating in similar utilities in neighboring towns and counties. Local people coming together to solve problems for the good of the community is how rural power got started, and it's how clean energy futures will get built.

Finally, as you're building and contributing to community, remember to make time together to share joy. Building clean energy futures can be hard work. It takes courageous innovation, and it can take extraordinary patience to work through the difficulties you will confront as you challenge old systems to accommodate new technologies and ideas. Developing friendships with people who share your vision makes the work lighter, and you will need them.

Building Momentum

We can now move forward with confidence that the shared services we will all need, like finance and training, are already in place at sufficient scale to support clean energy futures across rural America. This would not have been the case ten years ago. Back then, you would have had to invent many of tools and resources you needed on your own. Our task now is to build community, including the relationship infrastructure that will sustain us as individuals and leaders in our communities, and to develop our own customer-focused innovations that can also help meet other rural communities' needs.

Most people need to see something to believe it, so every success we each achieve within our community will help move us toward the futures we envision. And together, the collective impact of our achievements can build momentum toward a clean energy-powered transformation of rural America that rebuilds wealth, sustains jobs, conserves natural resources, protects the environment, and restores quality of life.

While individually, what we accomplish in service to our communities may seem small, our collective impact can make every hometown a place where families can thrive.

Conclusion
Sharing Power

We are in the midst of a technology-enabled transformation of our energy systems that can localize power using clean energy, revitalizing our hometowns. The tools and technologies we need are ready, and each has the potential to help repair our communities and restore justice in the process, but they won't implement themselves. The stories of local leadership throughout *Rural Renaissance* demonstrate just how much one person with conviction can do. That one person in your hometown could be you.

Your community needs you to help get the ball rolling to deploy energy efficiency, solar power, resilience, and electric vehicles—and to make sure it's all connected with accessible broadband. The foundations you need to start building your clean energy futures—financing models, workforce training, information systems and tools—are already in place with sufficient scale to support your work. Your first task is to make sure the values you build from are explicit, transparent, and right.

Systems produce outcomes according to the values on which they're founded. So, to change the outcomes of our energy systems, we've got to reform our values. This means reorienting our energy system away from a model that extracts resources, leaves pollution, concentrates wealth, and impoverishes communities and toward models that share power, share wealth, improve health, and sustain communities. Let's consider what it would look like to rebuild a localized clean energy system around the seven principles we discussed in the introduction.

Put the Public Back in Public Utilities

As we've learned, rural cooperatives and public power utilities were designed to be owned and controlled by the people they serve through governing bodies that are democratically elected by their customers. This revolutionary idea, which put customers in control instead of financial investors, originated in the nineteenth century, when farmers and workers were beginning to organize themselves to confront the low wages, miserable working conditions, and high prices of the day. By pooling their money, talents, and purchasing power, they were able to form cooperative businesses that paid better, treated workers with respect, and charged lower prices to participating members. The same self-organizing spirit led to the formation of cooperative and public power utilities.

Whether you're a member of the cooperative board or city council, an employee, or a member of the community, take a close look at how your local utility is operating. Is it living up to its revolutionary, democratic roots? Does its leadership represent the diversity of the community it serves? If not, now is a good time to put the public back in your local public utility. If it is operating as intended, your local utility will be better equipped to build a clean energy future that benefits everyone—though making decisions democratically is not always easy. People have diverse perspectives and often don't agree, and for good reasons, but debate and compromise can build consensus and lead to better-informed outcomes.

As we have seen, you have great models to follow. Curtis Wynn of SECO Energy has led the way to establish new national policies and programs on diversity, equity, and inclusion in cooperatives. And organizations like Kentuckians for the Commonwealth show us how to lead a grassroots reformation.

Align Value with Values

In concert with its governance model, the business model of your local nonprofit utility was designed to provide affordable electricity to everyone and to invest the resulting profits to support the development of your community. In today's business vernacular, your nonprofit utility functions as a social enterprise. Is your local utility living up to its nonprofit responsibility to operate for the benefit of the public good?

For-profit corporations are the dominant form used to organize business activities, and it's very easy for nonprofit institutions to fall into the habits of for-profit corporations simply because it's the cultural norm. For nonprofits, however, sometimes the right decision is to forgo revenue and profitability in favor of the mission if the two are in conflict. Is your local utility identifying and making those kinds of choices?

Nonprofits, including your local nonprofit utility, are responsible for measuring progress and success in terms of mission value. Value encompasses much more than money, and includes direct and indirect economic benefits, public health, happiness, and conservation of the natural environment. Money is just money. It's a useful tool, but money alone can't revitalize your community.

As you're getting to know your nonprofit utility's business model—how it earns revenue, where it gets its financing, its cost structure, and how it reinvests in the development of the community—take note of how the utility measures value, how it is connected to the values of your community, and how those values are represented in the utility's work. If your community values reducing poverty, is your utility delivering programs and services that help people living in poverty reduce their energy bills? If your community values conservation and the protection of the natural environment, has your utility committed to a specific clean energy goal? If the business model of your nonprofit utility does not reflect the values of your community, the current transformation of our energy system is an excellent opportunity to make some changes.

Repair, Restore, and Do Justice

The energy burden of spending more of your household income to keep the lights on than you do on groceries, or even rent, contributes to persistent rural poverty. As we've seen, while low-income households across America suffer with high utility bills because of old, inefficient housing, the impact is particularly acute in rural communities, where housing tends to be older and communities poorer. Moreover, energy burdens are disproportionately borne by Black and brown households, who also suffer the worst and first from the impacts of climate change and from the health impacts of fossil fuels. These racial disparities are not happenstance but the result of design, and they persist today as a part of the legacy of racist laws and policies like redlining, which denied mortgages and other types of financing to Black communities, and an energy system that created "sacrifice zones" where people's health and lives are lost to toxic pollution.

Energy efficiency is the most widely accessible type of clean energy, and you can use it to help repair, restore, and do justice in your community. Make decisions about your program's design that identify and prioritize the needs of the most energy-burdened people in your community, and be aware of the history of segregation in neighborhoods that continues to impact people and their housing. In addition to the energy efficiency improvements that you can make to lower people's bills, paid for with savings, organize local foundations and community service groups to help make home repairs that are necessary, yet outside the scope of energy efficiency. Patching or replacing leaky roofs, replacing rotted floorboards, and fixing leaky plumbing helps people and improves the quality and affordability of local housing—and you can do even more good by making sure that the repairs are completed by local tradespeople supported with technical workshops and training that welcome new, diverse colleagues into clean energy jobs.

Operating an energy efficiency program is also a wonderful opportunity for local contractors and your nonprofit utility to build new skills that they will need as your community moves forward into its clean energy future. From completing home energy assessments to managing an on-bill financing program, all the same capabilities will be used to expand access to solar power, increase resilience, and deploy electric vehicles.

Count Your Blessings and Work from Abundance

You can also choose to start building your clean energy future with solar power. The sun shines everywhere, and the land you need to install large-scale solar generation is abundantly available across rural America. In addition to installing solar on rooftops, businesses, and farms, we've also explored examples of how your local community can use solar as a part of an economic development strategy to attract businesses and new employers. Don't let anyone tell you it won't work in your state. Solar even has a place in remote Alaska.

Solar farms can work in concert with agriculture, including supporting pollinators or using agrivoltaic solar systems that pair solar panels with food crops to increase the yields of both. By being deliberate about where you develop solar projects and how they build local wealth, you can also protect multigenerational land ownership and Black-owned family farms. Remember Thomas Mitchell's scholarship and the loss of Black-owned land that continues today, and make sure that your state has protections in place that prevent unethical developers from forcing the sale of family land. One example is the Uniform Partition of Heirs Property Law, developed through Mitchell's leadership, which has already been passed in seventeen states.

State policy shapes possibility, and energy policy is very dynamic, so make sure you understand your state's energy policy environment and how it impacts the way you develop solar. Signing up for updates from Vote Solar, joining your state chapter of the Solar Energy Industries Association if you're in the energy business, or getting involved with Conservatives for Clean Energy are all helpful ways to connect and stay on top of your state's energy laws and regulations—and to get involved to change them if need be.

Share and Serve

As we know, local clean power—combined with energy storage and connected through local microgrids—increases energy resilience by enabling homes, neighborhoods, community centers, and other facilities to keep the power on even when the grid goes out. If we didn't have the technology to generate energy from renewable resources, local resilience would depend on diesel generators, electric vehicles would be powered by burning fossil fuels, and both would still contribute to the degradation of our health, well-being, infrastructure, and climate.

For many communities, perhaps even our own, increasing energy resilience addresses an urgent priority to respond to the present-day impacts of climate change. Historic events like the extreme cold in Texas in 2020 and recent fires in California highlight the urgent need for clean power generation and energy storage close to home to respond to our climate crisis.

Energy resilience does more than keep the power on, though. It helps to fulfill the potential of clean energy to localize our energy systems, enabling local energy systems to operate whether they are connected or disconnected from the larger electricity grid. The same technologies improve the efficiency of local energy systems, reduce infrastructure and operating costs, and may provide a more affordable and sustainable way of adding capacity to serve new electricity demand than building new substations.

Investing in energy resilience is an important opportunity to meet these needs and to share the benefits, like making sure that our most vulnerable neighbors are never without power. Corporate facilities like data centers can choose to build solar and energy storage systems that share resilience with facilities like hospitals, communities can build resilience centers in multifamily senior housing so that the elderly never face a power outage, and utilities can develop models for deploying resilience that share the potential cost savings with the community.

The energy technologies that come together to increase resilience and the business models that support their operation are still quite new, so focus first on a demonstration or pilot project that shows how increasing resilience with clean energy can share benefits across your community. The same rules apply to resilience as apply to solar: state and local policy will shape what's possible, and these policies may change as the market evolves. Take inspiration from the stories of people like Eric Clifton of Orison, who's building a new energy storage business to increase access to power and resilience, founded on his faith, from the small town where he lives in Wyoming.

Courageous Innovation

Similar to the technologies that use clean energy to increase energy resilience, the electrification of transportation is just beginning. As it builds momentum, it will transform our electric utilities and energy systems by shifting the energy consumption of cars, trucks, and other vehicles from fuel pumps to electricity. Instead of filling up at the gas station, we'll all plug in. As a result, other types of transportation infrastructure and services will change, too, including manufacturing, maintenance, and potentially even roadways.

Inspired by the courageous innovation of people like Rey León and organizations like The Ray, consider how your community's needs intersect with transportation. Is affordable transit a priority, or are there local auto manufacturers whose plants may need to retool? Are you near major distribution centers that will require access to electric vehicle charging infrastructure, or do you need to focus on affordable access for local residents?

Answering these and the many other questions we will face together as we electrify the transportation sector will shape your local clean energy future and may impact the course of many other communities too. Mayor León's EV transit model, for example, has grown to serve surrounding rural areas, and The Ray's innovations are being exported to other states. Whatever the path may be, it will vastly increase the revenue potential of your local nonprofit utility by increasing electricity demand and the array of transportation-related services they can offer. Make sure this additional value supports restorative investments and a thriving future for everyone in your community.

Partnership and Connection

Strong, values-aligned partnerships will be vital for electrifying the transportation sector in ways that revitalize your hometown. They are also essential to building connections between the broadband utility infrastructure that enables clean energy technologies to work together through a smart grid, and the rural residents and businesses that still need access to broadband services.

As we discussed earlier in this book, the same rural and small-town utilities that delivered power to farms in the 1930s are now building their own fiberoptic broadband networks to modernize their electricity distribution networks and create smart grids. They have the potential to extend their networks a little farther to provide broadband access to the 19 percent of rural households that lack it, thereby delivering universal access to high-speed internet just as they did for electricity. In the same ways that private electric utilities tried to stop them a hundred years ago, private telecommunications companies are blocking the expansion of rural broadband by cooperatives and public utilities today.

Shifting the conversation from protecting private profits by preventing competition to building partnerships that create connections for rural communities would be a game-changer. While advocating for better policies is a less concrete activity than installing a solar panel, it is of paramount importance for realizing the vision of a rural renaissance. Because high-speed broadband networks capable of handling vast quantities of data are necessary for a clean energy future, clean energy and our local nonprofit utilities can close rural America's broadband gap too.

Let's Get to Work

Energy is a building block of our society and economy, and the ways in which it is produced, distributed, and used are undergoing a massive transformation. The extent to which we take advantage of the opportunities this transformation presents to localize our energy systems and thereby drive investment into our communities will determine the extent to which the clean energy transformation will create a rural renaissance.

It's time to get to work. We've got the technologies and tools that we need, the freedom to define our own futures, the local utilities we need to build them, and great examples to follow. By connecting our efforts at the local level, we can change the landscape of opportunity for rural

America everywhere. By reforming the values at the heart of our energy systems in the process, we can also repair and restore our communities with energy by sharing power, sharing wealth, and protecting the places that we love. If we merely transform the technology but not the underlying systems and values, we will build a clean energy dystopia that widens existing wealth and health disparities.

It will be hard. Building a clean energy future takes innovation and creates change. There may be times when old ways of doing things and the people and companies that benefit from them resist you and fight for the status quo, and you may get tired. But as the Good Book says, "Let us not be weary in well doing: for in due season we shall reap, if we faint not."[1] So don't give up, keep going, and you'll see the results in our lifetimes.

Lead with love. Localizing your community's energy system with clean power is an opportunity to restore right relationships between the natural and the built environment and between energy and economic systems so that producing and delivering power doesn't extract wealth but rather sustains communities. If we keep the love we have for our hometowns and our neighbors at the forefront, it'll help us navigate the road ahead and make decisions we can be proud of.

Clean energy offers us a future; lots of different futures, in fact. The possibilities are as diverse as the places we live, the ways we make decisions, and the renewable resources we have to share. Let's use this gift, and this moment of transformation, to build the kinds of energy futures that will enable us all to thrive.

Epilogue
Going Home

Much like Samuel Pepys and his plague diary, which many of us remember from tenth-grade English Literature, I began writing this book at the onset of the 2020 COVID-19 pandemic. The November before, my husband and I had sold our home in Washington, DC, to settle together full-time in Midlothian, Virginia, about twenty minutes southwest of Richmond. Midlothian was the site of the first commercial coal mine in America, and it seemed like the perfect place to write about revitalizing your hometown around a clean energy future.

It was also the perfect place to live during a global pandemic. Midlothian was once the outermost suburb of Richmond, and even today, it's a short drive to the farmland of rural Virginia. Instead of being entirely shut in at our house, my husband and I and his young adult children were able to get outside every day for walks by the lake, along wooded paths, just by walking out our door.

As the pandemic wore on—and it still has not ended at the time I am writing this—changes in the US real estate market began to indicate a move back to small towns and to the country. We could see it south of the James River and in my hometown of LaGrange, Georgia, where suddenly homes with land out in the county were selling above asking price, sometimes in all cash, within days of being listed on the market.

The same trend was apparent in Gallup polling. In 2018, only 38 percent of Americans said they wanted to live in a rural community or a small town, but when asked the same question in 2020, the preference for living in the country had risen to 49 percent, with gains across every demographic.[1] In the realm of pop culture, HGTV shows like *Home*

Mammaw Moore's homeplace in Level Roads, Alabama, near where Pappaw Moore also grew up. I look forward to clean energy futures bringing new opportunities to their hometowns. (Moore family photo.)

Town, *Fixer to Fabulous*, and *Cheap Old Houses* presented affordable fantasies of small-town living, and *Fixer Upper*, which told stories about craftsmanship and home renovations in Waco, Texas, spawned its own media empire and cable network. It all made me wonder, were more Americans thinking about going back home? And if they did, how long would it last?

I wrote *Rural Renaissance* out of love for my hometown and for my neighbors, to share what I'd learned putting my experience in clean energy to work for my community. Maybe more people would be going back home and would share what they learned too?

Whether you live in your hometown, are going back, or are looking for a new hometown, I hope you can put the ideas and examples I've documented here to work to help all our rural communities become places where families can thrive for generations to come. Local clean energy systems, broadband included, give us all an opportunity to rebuild with better values so that we can have a healthier, more equitable, and ultimately more humane future.

As I've shared, my friend Ari Wallach would sum it up by telling us all to "be great ancestors," which is a humbling and very personal way to think about it. As we're considering how we can do just that—in how we build, share, and use clean energy, and in our everyday lives— let's commit ourselves to not allowing this incredible, transformative moment to pass us by.

Acknowledgments

I would like to praise God and express my gratitude as a Christian whose faith is central to my sense of purpose and why I do this work. I have prayerfully depended on God every day throughout the process of writing this book during 2020–21, which have been trying years for us all.

Notes

Prologue

1. Lydia Saad, "Country Living Enjoys Renewed Appeal in US," Gallup, January 5, 2021, https://news.gallup.com/poll/328268/country-living-enjoys-renewed -appeal.aspx.

2. Kim Parker et al., "What Unites and Divides Urban, Suburban, and Rural Communities," Pew Research Center, May 22, 2018, https://www.pewresearch .org/social-trends/2018/05/22/demographic-and-economic-trends-in-urban -suburban-and-rural-communities/; Eric Roston, "US Cities Are Under-Count ing Their CO_2 Pollution by Almost 20%," Bloomberg Green, February 22, 2021, https://www.bloomberg.com/news/articles/2021-02-02/u-s-cities-are-under -counting-their-co-pollution-by-almost-20.

3. Centers for Disease Control and Prevention, "Rural Americans Are Dying More Frequently from Preventable Causes Than Their Urban Counterparts," news release, November 7, 2019, https://www.cdc.gov/media/releases/2019/ p1107rural-americans.html.

4. Centers for Disease Control and Prevention, "CDC Reports Rising Rates of Drug Overdose Deaths in Rural Areas," news release, October 19, 2017, https:// www.cdc.gov/media/releases/2017/p1019-rural-overdose-deaths.html.

5. "181 Rural Hospital Closures Since 2005," Data Mapping Tool, Cecil G. Sheps Center for Health Services Research, University of North Carolina at Chapel Hill, accessed September 20, 2021, https://www.shepscenter.unc.edu /programs-projects/rural-health/rural-hospital-closures/.

6. Onyi Lam, Brian Broderick, and Skye Toor, "How Far Americans Live

from the Closest Hospital Differs by Community Type," Pew Research Center, December 12, 2018, https://www.pewresearch.org/fact-tank/2018/12/12/how-far-americans-live-from-the-closest-hospital-differs-by-community-type/.

7. "In Rural America, Too Few Roads Lead to College Success," *Focus*, Lumina Foundation, Fall 2019, https://focus.luminafoundation.org/in-rural-america-too-few-roads-lead-to-college-success/.

8. Sophia Campbell, Jimena Ruiz Castro, and David Wessel, "The Benefits and Costs of Broadband Expansion," *Up Front* (blog), Brookings Institution, August 18, 2021, https://www.brookings.edu/blog/up-front/2021/08/18/the-benefits-and-costs-of-broadband-expansion/.

9. Lichter and Ziliak, "The Rural-Urban Interface: New Patterns of Spatial Interdependence and Inequality in America," *ANNALS of the American Academy of Political and Social Science* 672, no. 1 (June 23, 2017): 6–25, https://doi.org/10.1177/0002716217714180.

Introduction

1. Matthew 6:21 (New Revised Standard Version).

2. Bartees Cox and Shravya Jain-Conti, "Pollution, Race, and the Search for Justice," Nexus Media News, March 1, 2018; https://nexusmedianews.com/pollution-race-and-the-search-for-justice-video-e962785d094d/.

Chapter 1

1. Will Wade and Eric Roston, "Getting US to Zero Carbon Will Take a $2.5 Trillion Investment by 2030," Bloomberg Green, December 15, 2020, https://www.bloomberg.com/news/articles/2020-12-15/getting-the-u-s-to-zero-carbon-would-cost-2-5-trillion-by-2030.

2. "2021 US Statistical Report," American Public Power Association, 2021, https://www.publicpower.org/system/files/documents/2021-Public-power-Statistical-Report.pdf.

3. *Final Report on the WPA Program, 1935–1943*, (Washington, DC: US Government Printing Office, December 18, 1946), iii–iv, http://lcweb2.loc.gov/service/gdc/scd0001/2008/20080212001fi/20080212001fi.pdf.

4. "Public vs. Private Power: From FDR to Today," *Frontline*, accessed September 20, 2021, https://www.pbs.org/wgbh/pages/frontline/shows/blackout/regulation/timeline.html#fn2.

5. Matt Novak, "How the 1920s Thought Electricity Would Transform Farms

Forever," *Gizmodo*, June 3, 2013, https://gizmodo.com/how-the-1920s-thought
-electricity-would-transform-farms-510917940.

6. "The Tennessee Valley Authority: Electricity for All," VCU Libraries: Social
History Project, Virginia Commonwealth University, accessed August 29, 2021,
https://socialwelfare.library.vcu.edu/eras/great-depression/tennessee-valley
-authority-electricity/.

7. Caterina Cowden, "Movie Attendance Has Been on a Dismal Decline since
the 1940s," *Business Insider*, January 6, 2015, https://www.businessinsider.com/
movie-attendance-over-the-years-2015-1.

8. "Joris Ivens," Great Directors, Senses of Cinema, accessed August 29, 2021,
https://www.sensesofcinema.com/2005/great-directors/ivens/.

9. Delia Patterson, "Public Power: A Rich History, A Bright Future," *Community Engagement* (blog), American Public Power Association, February 15, 2018,
https://www.publicpower.org/blog/public-power-rich-history-bright-future.

10. Martha Davis, "The Urge to Merge," T&D World, October 21, 2019,
https://www.tdworld.com/utility-business/article/20973336/the-urge-to-merge.

11. Megan Leonhardt, "Are You Paying Too Much for Your Phone? Here's How
to Potentially Save Hundreds per Year on Your Bill," CNBC, June 15, 2021, https:
//www.cnbc.com/2021/06/15/how-to-save-money-on-your-phone-bill.html;
Becky Chao and Claire Park, "The Cost of Connectivity 2020," New America,
July 15, 2020, https://www.newamerica.org/oti/reports/cost-connectivity-2020/
global-findings/; US Energy Information Administration, "Annual Electric Power
Industry Report," August 3, 2021, https://www.eia.gov/electricity/data/eia861/.

12. "Public Power," American Public Power Association, accessed September
20, 2021, https://www.publicpower.org/public-power.

13. "2021 Statistical Report," American Public Power Association, 2021, https:
//www.publicpower.org/system/files/documents/2021-Public-power-Statistical
-Report.pdf.

14. "The Economic Impact of America's Cooperatives," FTI Consulting,
prepared for the National Rural Electric Cooperative Association, March 25,
2019, https://www.fticonsulting.com/insights/reports/economic-impact-americas
-electric-cooperatives.

15. "Electric Cooperatives and Persistent Poverty Counties," *Business and
Technology Advisory*, National Rural Electric Cooperative Association, June 18,
2018, https://www.cooperative.com/programs-services/bts/Documents/Advisories
/Member-Advisory-on-Persistent-Poverty-Counties-June-2018.pdf.

Chapter 2

1. "Coal Blooded: Putting Profits before People," NAACP, accessed September 20, 2021, https://naacp.org/resources/coal-blooded-putting-profits-people.

2. American Council for an Energy Efficiency Economy, "Report: 'Energy Burden' on Low-Income, African American, and Latino Households Up to Three Times as High as Other Homes, More Energy Efficiency Needed," news release, April 20, 2016, https://www.aceee.org/press/2016/04/report-energy-burden-low-income.

3. Tom DiChristopher, "Gas Ban Monitor: States Launch Anti-Ban Blitz as Electrification Efforts Grow," News & Insights, S&P Global Market Intelligence, January 29, 2021, https://www.spglobal.com/marketintelligence/en/news-insights/latest-news-headlines/gas-ban-monitor-states-launch-anti-ban-blitz-as-electrification-efforts-grow-62336952.

4. Dennis Pillion, "State Lets Alabama Keep Solar Fee," AL.com, updated September 1, 2020, https://www.al.com/news/2020/09/state-lets-alabama-power-keep-solar-fee.html.

5. Notice of Adoption of New Fire Department Rule, 3 RCNY 608-01, titled "Outdoor Stationary Storage Battery Systems," New York City Fire Department, August 13, 2019, https://www1.nyc.gov/assets/fdny/downloads/pdf/codes/3-rcny-608-01.pdf.

6. "Ecoregions," US Environmental Protection Agency, accessed September 20, 2021, https://www.epa.gov/eco-research/ecoregions.

7. Greg A. Barron-Gafford et al., "Agrivoltaics Provide Mutual Benefits across the Food–Energy–Water Nexus in Drylands," *Nature Sustainability* 2 (2019): 848–55, https://doi.org/10.1038/s41893-019-0364-5.

Chapter 3

1. Jack Kittredge, "A History of the Cooperative Movement," *Natural Farmer*, Fall 2019, https://thenaturalfarmer.org/article/a-history-of-the-cooperative-movement/.

2. "Cooperative Identity, Values, and Principles," International Cooperative Alliance, accessed September 20, 2021, https://www.ica.coop/en/cooperatives/cooperative-identity.

3. "Cooperative Principles," Touchstone Energy Cooperatives, accessed September 20, 2021, https://www.touchstoneenergy.com/cooperative-principles.

4. Michael Seto and Cheryl Chastain, "General Survey of I.R.C. 501(c)(12) Cooperatives and Examination of Current Issues," Exempt Organizations Continuing Professional Education (CPE) Technical Instruction Program for Fiscal Year 2002, Internal Revenue Service, 2002, https://www.irs.gov/pub/irs-tege/eotopice02.pdf.

5. Tim Smart, "Who Owns Stocks in America? Mostly It's the Wealthy and White," *US News and World Report*, March 15, 2021, https://www.usnews.com/news/national-news/articles/2021-03-15/who-owns-stocks-in-america-mostly-its-the-wealthy-and-white.

6. John Farrel, Matt Grimley, and Nick Stumo-Langer, *Re-Member-ing the Cooperative Way*, Institute for Local Self-Reliance, March 2016, https://ilsr.org/wp-content/uploads/2016/03/Report_Remembering-the-Electric-Cooperative.pdf.

7. Abby Spinak, "Infrastructure and Agency: Rural Electric Cooperatives and the Fight for Economic Democracy in the United States," (PhD diss, MIT, 2014), 212–13, https://dspace.mit.edu/handle/1721.1/87519.

8. "Per Capita Income by Census Designated Place 2015–2019," South Carolina Census Data Center, South Carolina Revenue and Fiscal Affairs Office, accessed September 20, 2021, https://rfa.sc.gov/data-research/population-demographics/census-state-data-center/per-capita-income-place.

9. Avery Wilks, "SC Utility's Board Thrown Out of Office by More Than 1,500 Customers at Meeting, *State*, August 20, 2018, https://www.thestate.com/news/politics-government/article216754335.html.

10. Hunter Riggall, "LaGrange Settles Utilities Lawsuit Alleging Discrimination," *LaGrange Daily News*, November 2, 2020, https://www.lagrangenews.com/2020/11/02/lagrange-settles-utilities-lawsuit-alleging-discrimination/.

11. "The Crisis in Rural Electric Cooperatives in the South," Rural Power Project, May 6, 2016, https://ruralpowerproject.org/wp-content/uploads/2016/02/Rural-Power___Final.pdf; "From Power to Empowerment: Plugging Low Income Communities into the Clean Energy Economy," Groundswell, April 13, 2016, https://s3.amazonaws.com/groundswell-web-assets/documents/frompower_to_empowerment.pdf.

12. "Recent Trends in Poverty in the Appalachian Region," Applied Population Laboratory, University of Wisconsin, prepared for the Appalachian Regional Commission, August 2000, 89, https://www.arc.gov/wp-content/uploads/2020/06/RecentTrendsinPovertyintheAppalachianRegion.pdf.

13. Spinak, "Infrastructure and Agency," 200–202.

14. Home Page, Kentuckians for the Commonwealth, accessed September 20, 2021, https://kftc.org.

15. Spinak, "Infrastructure and Agency," 208–216.

16. "Democratizing Power in Rural America through Electric Co-Ops," *The Next System Project* (podcast), October 7, 2019, https://thenextsystem.org/learn/stories/democratizing-power-rural-america-through-electric-co-ops.

17. "Kentuckians for the Commonwealth Rural Electric Co-Op Reform Platform," Kentuckians for the Commonwealth, accessed September 20, 2021, https://archive.kftc.org/sites/default/files/docs/resources/coopreformplatformfinal.pdf.

18. Home Page, Rural Electric Cooperative Toolkit, accessed September 20, 2021, https://www.electriccooporganizing.org.

19. Algenon L. Cash, "Black History: A Conversation with Curtis Wynn," blog, Wharton Gladden, accessed August 29, 2021, https://whartongladden.com/black-history-conversation-curtis-wynn/.

20. Cathy Cash, "NRECA President Outlines Key to Co-Op Success in Industry Evolution," National Rural Electric Cooperative Association, March 13, 2019, https://www.cooperative.com/news/Pages/NRECA-President-Curtis-Wynn-Annual-Meeting.aspx.

21. "Success Stories," Rural Electric Cooperative Toolkit, accessed September 20, 2021, https://www.electriccooporganizing.org/success-stories.

Chapter 4

1. Seto and Chastain, "501(c)(12) Cooperatives"(see chap. 3, n. 4).

2. "What Is Public Power?," American Public Power Association, accessed September 20, 2021, https://www.publicpower.org/system/files/documents/municipalization-what_is_public_power.pdf.

3. Eric Hatlestad and Liz Veazey, "Commentary: Rural Power Co-ops 'Stranded in Coal,'" *Daily Yonder*, June 21, 2019, https://dailyyonder.com/commentary-rural-power-co-ops-stranded-coal/2019/06/21/.

4. Esther Whieldon and Gaurang Dholakia, "Forgiving Co-ops' Federal Coal Debt to Promote Renewables Faces Hurdles," News & Insights, S&P Global Market Intelligence, , October 9, 2019, https://www.spglobal.com/marketintelligence/en/news-insights/trending/TmCJLa6VEIt3_meORsYNyA2.

5. James Bruggers, "A Legacy of the New Deal, eElectric Cooperatives Struggle

to Make a Green Transition," *Courier-Journal*, March 1, 2021 (updated March 4, 2021), https://www.courier-journal.com/story/news/2021/03/01/rural-electric-co-ops-struggle-make-leap-greener-energy/6856700002/.

6. Randall Dupont, "The Role of Scientific Management in Rural Electrification: Morris L. Cooke's Quest for a Better Society," January 2, 2000, http://dx.doi.org/10.2139/ssrn.2711780.

7. "Municipal Bonds and Public Power," American Public Power Association, accessed September 20, 2021, https://www.publicpower.org/policy/municipal-bonds-and-public-power.

8. Daniel Bergstresser and Randolph Cohen, "Changing Patterns in Household Ownership of Municipal Debt: Evidence from the 1989–2013 Surveys of Consumer Finances," Abstract, Working Papers 87 (2015), Brandeis University, Department of Economics and International Business School, https://ideas.repec.org/p/brd/wpaper/87.html.

Chapter 5

1. Elvis Moleka, "The Poor Still Pay More: Solutions in Efficiency and Clean Energy for Energy Burdens & Climate Change," Groundswell, September 18, 2021, https://labs.groundswell.org/publications/.

2. Allison Plyer, "Facts for Features: Katrina Impact," Data Center, August 26, 2016, https://www.datacenterresearch.org/data-resources/katrina/facts-for-impact/.

3. Jennifer Ernst Beaudry, "Greening Up Employees," *Footwear News*, May 11, 2009, https://footwearnews.com/2009/business/news/sustainability-greening-up-employees-snow-socks-90430/.

4. "HEAL: Five Things to Know," Clinton Foundation, November 10, 2014, https://stories.clintonfoundation.org/heal-five-things-to-know-5f60d8f9e956.

5. "Opening Opportunities with Inclusive Financing for Energy Efficiency: Report on the First Year of the HELP PAYS Program at Ouachita Electric," Ouachita Electric Cooperative, June 2017, https://www.oecc.com/pdfs/HELP_PAYS_Report_2016-Ouachita_Electric_20170612V1.pdf.

6. "Creative Power: An Arkansas Electric Cooperative Changes the Game," *Groundtruth* (blog), Groundswell, October 30, 2017, https://groundtruth.groundswell.org/creative-power-how-an-arkansas-electric-cooperative-is-changing-the-game/.

7. Moleka, "The Poor Still Pay More."

8. Lauren Ross, Ariel Drehobl, and Brian Stickles, "The High Cost of Energy in Rural America: Household Energy Burdens and Opportunities for Energy Efficiency," American Council for an Energy Efficient Economy, July 2018, https://www.aceee.org/sites/default/files/publications/researchreports/u1806.pdf.

9. "LIHEAP Funding," Issues and Policy, Edison Electric Institute, accessed September 20, 2021, https://www.eei.org/issuesandpolicy/Pages/liheap.aspx.

10. "Life Expectancy: Could Where You Live Influence How Long You Live?" Robert Wood Johnson Foundation, accessed September 20, 2021, https://www.rwjf.org/en/library/interactives/whereyouliveaffectshowlongyoulive.html.

11. "Construction and Manufacturing Jobs Created per Million Dollars of Capital Investment in the Sustainable Recovery Plan," International Energy Agency, December 20, 2020, https://www.iea.org/data-and-statistics/charts/construction-and-manufacturing-jobs-created-per-million-dollars-of-capital-investment-in-the-sustainable-recovery-plan.

12. "Performance of Inclusive Financing for Energy Efficiency: Preliminary Results of the Ouachita Electric HELP PAYS Program," Ouachita Electric Cooperative, EEtility, Clean Energy Works, September 2016, http://www.oecc.com/pdfs/Ouachita%20Electric%20HELP%20PAYS%20Program%20-%20First%204%20Months%20of%20Activity.pdf.

13. "'Upgrade To $ave' Financed by USDA's Energy Efficiency and Conservation Loan Program" (PowerPoint presentation), National Academies of Sciences, Engineering, and Medicine, accessed September 20, 2021, https://sites.nationalacademies.org/cs/groups/depssite/documents/webpage/deps_170796.pdf.

14. Liam Jones, "Record $269.5bn Green Issuance for 2020: Late Surge Sees Pandemic Year Pip 2019 Total by $3bn," Climate Bonds Initiative, January 24, 2021, https://www.climatebonds.net/2021/01/record-2695bn-green-issuance-2020-late-surge-sees-pandemic-year-pip-2019-total-3bn.

Chapter 6

1. "Solar Industry Research Data," Solar Energy Industries Association, accessed September 20, 2021, https://www.seia.org/solar-industry-research-data.

2. "2021 Statistical Report," American Public Power Association, accessed September 20, 2021, https://www.publicpower.org/system/files/documents/2021-Public-power-Statistical-Report.pdf.

3. "North Carolina Solar and Agriculture," NC Sustainable Energy

Association, April 2017, https://energync.org/wp-content/uploads/2017/08/NCSEA_NC_Solar_and_Agriculture_7_27.pdf.

4. "Top 10 Solar States," Solar Energy Industries Association, accessed September 20, 2021, https://www.seia.org/research-resources/top-10-solar-states-0.

5. The White House, "Fact Sheet: Obama Administration Announced Additional Steps to Increase Energy Security," news release, April 11, 2012, https://obamawhitehouse.archives.gov/the-press-office/2012/04/11/fact-sheet-obama-administration-announces-additional-steps-increase-ener.

6. Walter C. Jones, "Solar Power Coming to Fort Gordon, Fort Stewart, Kings Bay, and Fort Benning," *Savannah Now*, October 23, 2014, https://www.savannahnow.com/news-latest-news/2014-10-23/solar-power-coming-fort-gordon-fort-stewart-kings-bay-and-fort-benning.

7. "Farms and Land in Farms, 2019 Summary," National Agricultural Statistics Survey, USDA, February 2020, https://www.nass.usda.gov/Publications/Todays_Reports/reports/fnlo0220.pdf.

8. Home page, Jack's Solar Garden, accessed September 20, 2021, https://www.jackssolargarden.com.

9. Georgena Terry, "State Pollinator-Friendly Solar Initiatives," Clean Energy States Alliance, December 2020, https://www.cesa.org/wp-content/uploads/State-Pollinator-Friendly-Solar-Initiatives.pdf.

10. Jared Brey, "Thomas W. Mitchell, MacArthur 'Genius' Grantee, on Black Landownership," *Backyard* (blog), Next City, October 27, 2020, https://nextcity.org/daily/entry/thomas-w-mitchell-macarthur-genius-grantee-on-black-land-ownership.

11. Emma Pollack and Fred Chung, "Clinic Students Reflect on 'Land Loss, Wealth, and Reparations,'" *CHLPI Blog*, Center for Health Law and Policy Innovation, January 27, 2020, Harvard Law School, https://www.chlpi.org/clinic-students-reflect-on-land-loss-wealth-and-reparations/.

12. Henry Louis Gates Jr., "The Truth Behind 40 Acres and a Mule," PBS, originally posted on *The Root*, accessed September 20, 2021, https://www.pbs.org/wnet/african-americans-many-rivers-to-cross/history/the-truth-behind-40-acres-and-a-mule/.

13. Greg Barlow, "Defying the Odds-Mitigating Property Loss Through Historic Partition Law Reform in the US," News and Press, LSA Stories, Law & Society Association, August 26, 2020, https://lawandsociety.site-ym.com/news

/523353/Defying-Great-Odds--Mitigating-Property-Loss-Through-Historic
-Partition-Law-Reform-in-the-U.S..htm.

14. "SUNDA Project: Helping Cooperatives Accelerate Deployment of Solar
PV," National Rural Electric Cooperative Association, accessed September 20,
2020, https://www.cooperative.com/programs-services/bts/sunda-solar/Pages
/default.aspx; "LIFT Solar: Financial Innovation to Accelerate Solar Access,"
Groundswell, accessed September 20, 2021, https://labs.groundswell.org/research
/#lift-solar-financial-innovation-to-accelerate-solar-access.

15. Darrell Proctor, "Corporate Clean Energy Purchases hit record," *Power*,
January 16, 2021, https://www.powermag.com/corporate-clean-energy-purchases
-hit-record/.

16. "Our Vision," Renewable Energy Buyers Association, accessed September
20, 2021, https://rebuyers.org/about/vision/.

17. "Electric Cities Thrive in Georgia Thanks to Senate Bill 95," Conservatives
for Clean Energy Georgia, April 22, 2019, https://www.cleanenergyconservatives
.com/electric-cities-thrive-in-georgia-thanks-to-senate-bill-95/.

18. "Google Building $600 million Data Center on Former TVA Coal Plant,"
News, Tennessee Solar Energy Association, April 11, 2018, https://www.tnsolar
energy.org/home/2018/4/11/google-building-600-million-data-center-on-former
-tva-coal-plant.

19. "More than a Megawatt: Embedding Social and Environmental Impact in
the Renewable Energy Procurement Process," Salesforce, June 27, 2019, https:
//www.salesforce.com/content/dam/web/en_us/www/assets/pdf/sustainability/
sustainability-more-than-megawatt.pdf.

20. "Corporates and Communities," Renewable Energy Buyers Alliance,
accessed September 20, 2021, https://reba-institute.org/programs/beyond-the
-megawatt/corporates-and-communities/dt/section/3/step/4/.

21. "The Community Solar Playbook," Toolkits and Samples, National Rural
Electric Cooperative Association, September 9, 2016, https://www.cooperative
.com/programs-services/bts/Pages/SUNDA/The-Community-Solar-Playbook.aspx.

22. Erik Daily, "Vernon Electric Cooperative Unveils State's First Community
Solar Project," *La Crosse Daily*, June 26, 2014, https://madison.com/business/
vernon-electric-cooperative-unveils-state-s-first-community-solar-project/article
_fdf7a25d-f431-5fdb-8d84-d552fca09dc1.html.

23. "Let the Sun Shine: 10 Things You Should Know About Community

Solar," blog, Walton EMC, June 17, 2019, https://www.waltonemc.com/index.php/blog/2019/06/let-the-sun-shine-10-things-you-should-know-about-community-solar.

24. Barbara Bradford, "Let There Be Light: Anza Electric Cooperative, The History," Anza Electric Cooperative, accessed September 20, 2021, http://anzaelectric.coopwebbuilder2.com/sites/anzaelectricanzaelectric/files/PDF/let_there_be_light_booklet_0.pdf.

25. "ACCESS Project Case Study: Anza Electric Cooperative," National Rural Electric Cooperative Association, October 2020, https://www.cooperative.com/programs-services/bts/access/Documents/Advisory-ACCESS-Case-Study-Anza-Oct-2020.pdf.

26. Christine Patmon, "PowerFin Brings SolarHost, First-of-Its-Kind, Rooftop Solar Program to San Antonio," CPS Energy, September 2, 2015, http://newsroom.cpsenergy.com/powerfin-brings-solarhost-first-of-its-kind-rooftop-solar-program-to-san-antonio/.

27. "Solar Power Q&A," Residential Solar, Ouachita Electric Cooperative Corporation, accessed September 20, 2021, https://www.oecc.com/solar.

Chapter 7

1. Ula Chrobak, "The US Has More Power Outages Than Any Other Developed Country. Here's Why," *Popular Science*, August 17, 2020, https://www.popsci.com/story/environment/why-us-lose-power-storms/.

2. Rebecca Smith, "How a U.S. Utility Got Hacked," *Wall Street Journal*, December 20, 2016, https://www.wsj.com/articles/how-a-u-s-utility-got-hacked-1483120856.

3. David Sanger and Nicole Perlroth, "Pipeline Attack Yields Urgent Lessons about U.S. Cybersecurity," *New York Times*, May 14, 2021, https://www.nytimes.com/2021/05/14/us/politics/pipeline-hack.html.

4. Lynn Doan, "How Many Millions Are without Power In Texas? It's Impossible to Know for Sure," *Time*, February 17, 2021, https://time.com/5940232/millions-without-power-texas/.

5. Scott Carpenter, "Traffic to Rooftop Solar Website Surges as Texans Seek Freedom," *Forbes*, February 18, 2021, https://www.forbes.com/sites/scottcarpenter/2021/02/18/traffic-to-rooftop-solar-sites-jumps-as-texans-seek-freedom-from-grid/?sh=2dac7fe8691d.

6. Stuart C. Gilson and Sarah L. Abbott, "PG&E and the First Climate Change Bankruptcy," Abstract, Harvard Business School Case 221-057, December 2020, https://store.hbr.org/product/pg-e-and-the-first-climate-change-bankruptcy/221057?sku=221057-PDF-ENG.

7. "Energy and Water Use in California are Interconnected," Energy and Water, Public Policy Institute of California, October 2016, https://www.ppic.org/wp-content/uploads/R_1016AER.pdf.

8. Nicolas C. Abi-Samra, "One Year Later: Superstorm Sandy Underscores Need for a Resilient Grid," *IEEE Spectrum*, November 4, 2013, https://spectrum.ieee.org/one-year-later-superstorm-sandy-underscores-need-for-a-resilient-grid#toggle-gdpr.

9. "Microgrids," Grid Modernization, National Renewable Energy Laboratory, accessed September 20, 2021, https://www.nrel.gov/grid/microgrids.html.

10. Jodan Wirfs-Brock, "Lost in Transmission: How Much Electricity Disappears between a Power Plant and Your Plug," *Inside Energy*, November 6, 2015, http://insideenergy.org/2015/11/06/lost-in-transmission-how-much-electricity-disappears-between-a-power-plant-and-your-plug/.

11. Elisa Wood, "What Is a Microgrid," *Microgrid Knowledge*, March 28, 2020, https://microgridknowledge.com/microgrid-defined/.

12. Phoebe Wall Howard, "Ford F-150 Generator Saves Farmington Hills Wedding Reception during Power Outage," *Detroit Free Press*, August 18, 2021, https://www.freep.com/story/money/cars/ford/2021/08/18/ford-f-150-generator-power-outage-wedding-reception/8156391002/.

13. Dennis Washington, "Alabama Power's Smart Neighborhood Wins Award," *Alabama News Center*, June 3, 2020, https://alabamanewscenter.com/2020/06/03/alabama-powers-smart-neighborhood-wins-award/.

14. "The Baltimore City Community Resiliency Hub Program," Baltimore Office of Sustainability, accessed September 20, 2021, https://www.baltimoresustainability.org/baltimore-resiliency-hub-program/.

15. Cameron Oglesby, "The Navajo Nation Generates a Ton of Power—but 14,000 Homes Don't Have Electricity," *Grist*, March 5, 2021, https://grist.org/justice/navajo-nation-electricity-power-covid/.

16. "Weather Related Fatality and Injury Statistics," Weather.gov, accessed September 20, 2021, https://www.weather.gov/hazstat/.

17. Cathy Cash, "Texas Co-Op Enters Energy Storage Arena with Grid-Scale Battery System," NRERCA, October 6, 2020, https://www.electric.coop/texas-co-op-enters-energy-storage-arena-with-grid-scale-battery-system.

18. Titiaan Palazzi and Dan Seif, "Why Distributed Solar Is Winning in Texas," *GreenBiz*, February 28, 2018, https://www.greenbiz.com/article/why -distributed-solar-winning-texas.

19. Matthew 25:40 (King James Version).

Chapter 8

1. "Energy Use for Transportation," Energy Use Explained, US Energy Information Administration, accessed September 20, 2021, https://www.eia.gov/ energyexplained/use-of-energy/transportation-in-depth.php.

2. Johnsen Del Rosario, "Bringing Clean Transportation Service to Huron, CA," Latest News, UpLiftCA, August 12, 2019, https://upliftca.org/portfolio/rey -leon/.

3. "Case Study: The Story of Green Raiteros: A Shared and Electric Lifeline for California Farmworkers, 2020," Share Use Mobility Center, 2020, https:// learn.sharedusemobilitycenter.org/casestudy/the-story-of-green-raiteros-a-shared -electric-lifeline-for-california-farmworkers-2020/.

4. María G. Ortiz-Briones, "LEAP's Green Raiteros Electric Vehicle Ridesharing Program Gets Loaners to Expand Services," *Fresno Bee*, August 12, 2021, https://www.fresnobee.com/vida-en-el-valle/noticias/california-es/fresno/article 253432099.html.

5. Britta Gross, "How to Move America to Electric Vehicles," *RMI Outlet* (blog), Rocky Mountain Institute, January 28, 2021, https://rmi.org/how-to -move-america-to-electric-vehicles/.

6. The White House, "Fact Sheet: Biden Administration Advances Electric Vehicle Charging Infrastructure," news release, April 22, 2021, https://www .whitehouse.gov/briefing-room/statements-releases/2021/04/22/fact-sheet-biden -administration-advances-electric-vehicle-charging-infrastructure/.

7. Alex Brown, "Electric Cars Will Challenge State Power Grids," *Stateline* (blog), Pew Charitable Trusts, January 9, 2020, https://www.pewtrusts.org/en/ research-and-analysis/blogs/stateline/2020/01/09/electric-cars-will-challenge-state -power-grids.

8. Anh Bui, Nic Lutsey, and Peter Slowik, "Briefing: Power Play: Evaluating the US Position in the Global Electric Vehicle Transition," International Council on Clean Transportation, June 29, 2021, https://theicct.org/publications/us -position-global-ev-jun2021.

9. Erin Schilling, "SK Innovation's Georgia Battery Plant to Employ 1,000 Workers by End of the Year," *Atlanta Business Chronicle*, April 21, 2021, https:/

/www.bizjournals.com/atlanta/news/2021/04/21/sk-innovation-georgia-battery
-plant-hiring.html.

10. Christian Spencer, "Finally, Here's the Exact Cost of Owning an Electric
Car vs. Gas Car," *Hill*, June 23, 2021, https://thehill.com/changing-america/sus
tainability/energy/559971-finally-heres-the-exact-cost-of-owning-an-electric-car.

11. "Plug In to Your Future—Drive Electric!" New Hampshire Electric Coop,
accessed September 20, 2021, https://www.nhec.com/drive-electric/.

12. "Response Bonanza for Utilities, If They Can Handle It," *Utility Dive*,
Industry Dive, October 18, 2019, https://www.utilitydive.com/news/ev-charging
-promises-a-demand-response-bonanza-for-utilities-if-they-can-h/563453/.

13. Utah Inland Port Authority, "UIPA and Rock Mountain Power Announce
Clean Energy Cooperation Agreement," news release, May 13, 2020, https:/
/inlandportauthority.utah.gov/2020/05/13/uipa-and-rocky-mountain-power
-announce-clean-energy-cooperation-agreement/.

14. Skip Descant, "In a Maryland County, the Yellow School Bus Is Going
Green," *Government Technology*, June 17, 2021, https://www.govtech.com/fs/in-a
-maryland-county-the-yellow-school-bus-is-going-green.

15. Tennessee Valley Authority, "Electric Vehicle Fast Charging Network Coming in Tennessee," news release, February 3, 2021, https://www.tva.com/news
room/press-releases/electric-vehicle-fast-charging-network-coming-to-tennessee.

16. Tennessee Valley Authority, "TVA Board Approves First Steps in Electric
Vehicle Infrastructure Initiative," news release, November 13, 2020, https://
www.tva.com/newsroom/press-releases/tva-board-approves-first-steps-in-electric
-vehicle-infrastructure-initiative.

17. Francis Energy, "Francis Energy Brings Fast Charging Capability to Cherokee Nation for Its First Electric Buses," news release, May 13, 2021, https://www
.prnewswire.com/news-releases/francis-energy-brings-fast-charging-capability-to
-cherokee-nation-for-its-first-electric-buses-301291166.html.

Chapter 9

1. David Thill, "Smart Grid Transformation Hinges on Data Bandwidth—and
Lots of It," *Energy News Network*, June 10, 2019, https://energynews.us/2019/06
/10/smart-grid-transformation-hinges-on-data-bandwidth-and-lots-of-it/.

2. Campbell, Castro, and Wessel, "Broadband Expansion" (see prologue, n. 8).

3. John B. Horrigan, "Baltimore's Digital Divide: Gaps in Internet Connectivity and the Impact on Low-Income City Residents," Abell Foundation, May

2020, https://abell.org/sites/default/files/files/2020_Abell_digital%20divide
_full%20report_FINAL_web%20(dr).pdf.

4. Campbell, Castro, and Wessel, "Broadband Expansion."

5. Sean Gonsalves, "The Problem(s) of Broadband in America," Institute for
Local Self-Reliance, July 2021, https://ilsr.org/wp-content/uploads/2021/07/
Problems-of-Broadband-072021.pdf.

6. Campbell, Castro, and Wessel, "Broadband Expansion."

7. Campbell, Castro, and Wessel.

8. Tyler Cooper, "Municipal Broadband Is Restricted in 18 States across the
US in 2021," Broadband Now, April 6, 2021, https://broadbandnow.com/report/
municipal-broadband-roadblocks/.

9. Editors (Stateline.org), "State Laws Slow Down High Speed Internet for
Rural America," *Government Technology*, January 14, 2019, https://www.govtech
.com/network/state-laws-slow-down-high-speed-internet-for-rural-america.html.

10. Doug Dawson, "The Rural Broadband Industry," CCG Consulting for the
Pew Charitable Trusts, September 1, 2021, https://www.pewtrusts.org/-/media/
assets/2021/09/white_paper_rural_broadband_industry_final.pdf.

11. Karen Haywood Queen, "In Virginia, Utility Smart Grid Projects Could
Fill Rural Broadband Gaps," *Energy News Network*, February 8, 2021, https://
energynews.us/2021/02/08/in-virginia-utility-smart-grid-projects-could-help-fill
-rural-broadband-gaps/.

12. "Dominion Energy Virginia: A Middle Provider Awaits Regulatory OKs
on Proposed Pilots," *Daily Energy Inside*, November 12, 2020, https://daily
energyinsider.com/policy/27977-dominion-energy-virginia-a-middle-mile
-provider-awaits-regulatory-oks-on-proposed-pilots/.

13. "Georgia's Approach to Rural Broadband," Georgia Technology Authority,
accessed September 20, 2021. https://gta.georgia.gov/georgias-approach-rural
-broadband1.

14. "Diverse Power Approved for Retail Broadband Affiliate," A Word from
Wayne, News from Diverse Power, June 3, 2020, http://www.diversepower.com/a
-word-from-wayne-diverse-power-approved-fro-retail-broadband-affiliate/.

15. Dawson, "The Rural Broadband Industry."

Chapter 10

1. Liam Jones, "$1Trillion Mark Reached in Global Cumulative Green Issu-
ance: Climate Bonds Data Intelligence Reports: Latest Figures," Climate Bonds

Initiative, December 15, 2020, https://www.climatebonds.net/2020/12/1trillion -mark-reached-global-cumulative-green-issuance-climate-bonds-data-intelligence.

2. "US Solar Industry Diversity Study 2019," Solar Foundation and the Solar Energy Industries Association, 2019, https://irecusa.org/wp-content/uploads /2021/07/Solar-Industry-Diversity-Study-2019-2.pdf.

Conclusion
1. Galatians 6:9 (King James Version).

Epilogue
1. Saad, "Country Living" (see prologue, n. 1).

Index

About the Author

Michelle Moore with her husband, Linwood Boswell.

L. Michelle Moore is CEO of Groundswell, a nonprofit organization that builds community power by connecting solar, resilience, and energy efficiency with economic development, affordability, and quality of life. A social entrepreneur and former White House official with roots in rural Georgia, Michelle has been helping communities across America shape sustainable, clean energy futures for twenty-five years. Her accomplishments range from leading the effort to cut federal energy use by $11 billion and deploying 3 gigawatts of new renewable energy for President Obama to developing innovative new clean energy programs for her hometown. Michelle was born and raised in LaGrange, Georgia, and now lives in Midlothian, Virginia, with her family. Her work is rooted in her faith, and in the commandment to "love your neighbor as yourself."